D0975460

BODIES, COMMODITIES, AND BIOTECHNOLOGIES

The Leonard Hastings Schoff Memorial Lectures

UNIVERSITY SEMINARS

Leonard Hastings Schoff Memorial Lectures

The University Seminars at Columbia University sponsor an annual series of lectures, with the support of the Leonard Hastings Schoff and Suzanne Levick Schoff Memorial Fund. A member of the Columbia faculty is invited to deliver before a general audience three lectures on a topic of his or her choosing. Columbia University Press publishes the lectures.

David Cannadine, *The Rise and Fall of Class in Britain* 1993

Charles Larmore, *The Romantic Legacy* 1994

Saskia Sassen, *Sovereignty Transformed: States and the New Transnational Actors* 1995

Robert Pollack, *The Faith of Biology and the Biology of Faith: Order, Meaning, and Free Will in Modern Medical Science* 2000

Ira Katznelson, *Desolation and Enlightenment: Political Knowledge After the Holocaust, Totalitarianism, and Total War* 2003

Lisa Anderson, *Pursuing Truth, Exercising Power: Social Science and Public Policy in the Twenty-first Century* 2003

Partha Chatterjee, *The Politics of the Governed: Reflections on Popular Politics in Most of the World* 2004

David Rosand, *The Invention of Painting in America* 2004

BODIES, COMMODITIES, AND BIOTECHNOLOGIES

Death, Mourning, and
Scientific Desire in the Realm
of Human Organ Transfer

LESLEY A. SHARP

Columbia University Press New York

COLUMBIA UNIVERSITY PRESS
Publishers Since 1893
New York Chichester, West Sussex

Library of Congress Cataloging-in-Publication Data

Sharp, Lesley Alexandra.
Bodies, commodities, and biotechnologies : death, mourning, and scientific desire in the realm of human organ transfer / Lesley A. Sharp.
p. cm. — (Leonard Hastings Schoff memorial lectures)
Includes bibliographical references and index.
ISBN 0-231-13838-5 (cloth : alk. paper) — ISBN 0-231-51098-5 (electronic)
1. Organs (Anatomy)—Social aspects—United States.
2. Transplantation of organs, tissues, etc.—Social aspects—United States.
3. Body, Human—Social aspects—United States.
4. Body, Human—Symbolic aspects—United States.
I. Title. II. University seminars/Leonard Hastings Schoff memorial lectures.
GT497.U6S53 2007
306.4—dc22
2006001648

Columbia University Press books are printed
on permanent and durable acid-free paper.
This book was printed on paper with recycled content.

Printed in the United States of America

c 10 9 8 7 6 5 4 3 2 1

For Erik,
my wonderful brother

CONTENTS

ILLUSTRATIONS

ACKNOWLEDGMENTS

A host of talented, helpful, and supportive people have made this experience truly pleasurable. First and foremost I thank Robert Belknap, Director of the University Seminars, who tracked me down when I was on sabbatical leave in order to extend this invitation. His persistent enthusiasm for this project's focus was a strong source of encouragement as I prepared and delivered these three talks. Amanda Roberts and Alison Garforth were outstanding in every way, ensuring not only that these events ran smoothly but also that they were beautifully catered and staged. Our afternoon spent over the tasting menu is especially memorable! The exceptional talents of Kari Hodges and Scott Michener enabled me to weave a range of visual materials into the first and third of my talks and, I believe, to stave off audience fatigue during lectures that followed lavish dinners. It was a special honor to have such illustrious colleagues as Jack Hawley, Richard Parker, and Judith Shapiro present to introduce the three lectures. The fact that at the time Richard was miserable from severe back pain underscores how exceptional a person he is. I also wish to thank my editor, Anne Routon, whom I fear I would never have met were it not for this speaking engagement. It is rare that an anthropologist has an opportunity to work with an editor who thinks like an anthropologist, too. Leslie Kriesel likewise proved to be a delightful manuscript editor, and the artist, Martha Lewis, and designer, Linda Secondari, captured the theme of the book perfectly with the jacket illustration and layout. Perceptive comments from two reviewers also proved of great help as I revised the lectures for publication.

A range of individuals were particularly helpful to my search for images to illustrate key points. I am especially indebted to the staff and members of the National Donor Family Council, especially those who have worked directly on the Patches of Love Quilt Project; Krissy Giacoletto and Roy Webb of the J. Willard Marriott Library at the University of Utah, Salt Lake City, whose determination to locate photographs made me realize what a treasure hunt this could be; and Robert Dicks of the Lucile Packard Children's Hospital at Stanford University, who, even amid another ground-breaking media event, somehow found the time to assist me in my endeavors. Through him, I wish to thank the family of Miles Coulson, too.

Together my husband and son, Andy and Alex Fox, did a marvelous job of keeping me in good humor throughout the month of November when I delivered these three talks, while Lindsey Owen freed me in wonderful ways from my obligations at home. Finally, I offer a special word of gratitude to the late Morton Klass, who was an invaluable mentor throughout my early years at Barnard College, and who first introduced me to the University Seminars at Columbia University. As noted during my original presentation, the first lecture—and, now, chapter—is dedicated in his honor.

BODIES, COMMODITIES, AND BIOTECHNOLOGIES

INTRODUCTION

Solitude is a cherished aspect of quotidian life for scholars who work within the academy. Were it not for the classroom, the college or university professor might well assume that an interested public extends only as far as the limits of his or her own discipline, limits readily apparent during regularly attended yet highly specialized conferences or seminars. The invitation to deliver the Leonard Hastings Schoff Memorial Lectures, then, has been a double blessing for me. First, I consider it an extraordinary honor to be invited to present this series of three talks; second, doing so has tested my ability to transform often obscure academic speech into palatable form that might be of interest to a wider audience. Although my current research focuses on a rarefied realm of medicine, I am most concerned with the daily experiences of patients and family members. I therefore commit an injustice against them if I describe their lives exclusively in a manner that only specialists can understand. The Schoff lecture format offers an important opportunity to speak plainly and directly about events that simultaneously mark mundane, remarkable, and bizarre aspects of life in America. Finally, this particular venue offers yet another luxury: that of playfulness. I hope that my audiences have enjoyed these presentations at least half as much as I have enjoyed delivering them.

These lectures are framed by my interest as an anthropologist in the increased commodification of the human body and its parts, a trend that figures in many chapters of human history and cultural settings but has become an especially pervasive—not to mention pernicious—force specifically within the United States in the late twentieth and early twenty-first centuries. The concerns outlined in this work have been shaped by my long-term ethnographic involvement within the realm of "organ transfer," an expression I employ in order to underscore the interdependence of organ transplantation, procurement, and donation, especially in this country. I am indebted to Shirley Lindenbaum for inspiring me to probe other quarters of life, too, a process that began when she requested that I write an essay on "The Commodification of the Body and Its Parts" (Sharp 2000).

As I have argued throughout my three lectures and within the chapters published here, organ transfer is fraught with hidden forms of body commodification. The ideological principles that guide acceptable practices, however, render commodification an elusive trend within the American[1] context. As an anthropologist, I have long been fascinated by the inevitable disjunction between the rules that supposedly govern human behavior and the regularity with which members of a particular group or community ignore or circumvent accepted regulations. It is a well-recognized social fact within the discipline that defiance is normative in human societies. Put simply, in some contexts it is impossible to follow all the rules all the time; in others, one might even argue that rules were made to be broken. But whereas flexibility is part and parcel of many social systems, within the context of my own research I regularly encounter an altogether different set of beliefs and behaviors at work.

Human action within the highly technocratic (Davis-Floyd 1994) medical realm of organ transfer is guided by strict rules of decorum. Such rules are framed by a sentimental structure of sorts: they privilege altruistic behavior and condemn marketing strategies of any kind. Thus, transplantable organs are always described as being donated by individuals or, if the individuals are deceased, by their surviving kin. The bodies from which such organs are taken are also always referred to as "donors," rather than "patients," or, even worse, "corpses" or "cadavers." The selflessness embedded (or, better yet, embodied) in such offerings forbids all mention of the marketing of the body. Because such offerings are made anonymously, the act of giving also insists that the giver let go once and for all, and, further, expect no reciprocal act of compensation.

Whereas involved health care professionals often do not question such premises, lay participants—most notably transplant recipients and the surviving kin of deceased donors—often struggle to make sense of the peculiarities of this gift economy. An especially striking response involves subversive acts designed to commemorate organ donors as a highly specialized and beloved category of lost dead. As such, these lectures only partly concern the unintended outcomes of a dialectic involving, on the one hand, biotechnological inventiveness and, on the other, the medical use of the human body. Inevitably, my discussions must also focus on death or, more specifically, on collective human responses when dead bodies are put to good use—to extend lives, to relieve suffering, or even to assuage guilt.

The anthropological literature teems with more than a century's worth of work dedicated to beliefs about and rituals surrounding death. In essence, then, my own focus fits squarely within the discipline, albeit with a special twist: I am especially concerned with the social responses among parties who are deeply involved in this specialized medical realm, a realm predicated on denying the commodification of the body and even the presence of the dead body itself. As we shall see, the all-too-perplexing ironies at work are exposed through a specific line of questioning. First, how are we to mourn organ donors when their deaths have been so carefully orchestrated by medical personnel? How are we to speak of their bodies when their parts have been offered to other strangers living in diverse sectors of the nation? And how are we to understand donors' deaths (or futures) when parts of them, at least, seem to thrive within the bodies of others?

With these initial questions in mind, I offer one final reflection on the peculiar nature of the medicalized body before I turn to the first chapter of this work. The questions I raise are drawn from responses offered by interviewees over the course of close to fifteen years of research. Many of the quandaries they voice directly concern the integrity of the body, especially when the body is estranged from its most essential parts. At issue here too are perplexing questions of personhood. These are evident in worries that focus on the destiny of a donor's soul, or whether organ recipients sense that others now reside within them.

As I will illustrate at the end of this book, the question of personhood similarly haunts experimental realms of research, where scientists imagine that relatively soon we may eliminate the need for human donors altogether in favor of parts of mechanical or animal origin. In so doing, we may well bypass those troubling questions about death and dead (human) bodies. As an anthropologist, I predict, however, that we will nevertheless be faced with still another set of ethical questions concerning the medical use of bodies and their parts, as well as the limits of personhood and, ultimately, human nature.

On that note, I turn to the lectures themselves, where I seek to untangle the troublesome themes of the good death, of body commodification, and of experimental (and, currently, futuristic) realms of biomedical research, all set against the intriguing backdrop of organ transfer in America.

ONE

THE GOOD DEATH

Managing and Memorializing the Dead

OPENING REFLECTIONS

Human bodies are fascinating things, their morphology and capabilities central to scientific definitions of what or who we are as a species in both contemporary and evolutionary terms. In this society, we nevertheless consider each body to be unique: even when confronted with what we call "identical" twins, we make special effort to distinguish one from the other. We embellish our bodies, too, with elaborate forms of clothing and other accoutrements, and we paint, pierce, and reshape everything from faces and earlobes to teeth, chins, and buttocks. We tone the body and mold it through exercise and surgery. We display who we are through our bodies; in this sense, we can speak of an "embodied" self. In many ways, we *are* our bodies. Thus, we might think of the body's outward appearance as a social skin of sorts (Bourdieu 1994 [1977]; Csordas 1994; Mauss 1973 [1935]; Scheper-Hughes and Lock 1987).

A paradox of the embodied self is that we regularly take it for granted. When the body functions properly, we rarely, if ever, pause to think about how we move through the world while assuming a bipedal stance, or, perhaps, about the ease with we reach for a glass, eating utensil, or other tool, and then pick it up with the aid of a uniquely human opposable thumb. Instead, we typically forget the body—our individual body—except under circumstances when it fails. As philosopher Drew Leder asserts, our bodies are typically "absent" from our consciousness when we are healthy (Leder 1990). When the body fails us, we suddenly become acutely—and painfully—aware of its presence and our dependence on it (cf. Merleau-Ponty 1962; Murphy 1987; Sacks 1985, 1998).

But let us push this concept farther, from the vital body through the terrain of the dysfunctional body and on to the dead body. How do we think about corpses? Or, perhaps, how often do we think about corpses? All the time, you might say, if you are a pathologist, homicide detective, or forensic anthropologist. But for most of us, I believe I am safe in assuming, the answer is not much. I would like to consider this question seriously, though: how much of us remains "in there," so to speak, once we are dead? Can we speak of the dead as possessing selfhood? Or, better put, perhaps (because selfhood implies consciousness), do we understand the dead as still possessing a social skin? Do the dead, at the very least, still bear some social value for us?

Granted, these are bizarre questions on the face of it. This is because in American society we spend very little time talking about the dead. As myriad examples from fiction and film attest, we fear bodies, especially when they are decaying corpses. To talk of death is repulsive; in doing so we raise the specter of the *memento more*. Our aversion stems, too, from the assumption that dead bodies are sources of death for others, harboring diseases that can infect and kill. Corpses are unclean, and thus bear the threat of both very real and imagined or symbolic forms of contagion. Finally, their presence makes us profoundly sad.

In response, we hide the dead from plain sight. A century ago a regular aspect of daily life involved bearing witness to the deaths of others: failing bodies were managed with care by close family members, religious officials, and physicians who made house calls. As John Berger documented nearly forty years ago in his book *A Fortunate Man*, the witnessing of life and death together similarly defined intimate aspects of work for an English country doctor (Berger 1967). Today in America, however, we readily shuttle away the dying to hospitals and hospices so that their failing bodies can be managed with care by specialists, or what David Rothman has referred to as the "strangers at the bedside" (Rothman 1991). These specialists are physicians and nurses trained to keep their patients alive, not manage their deaths (Nuland 1993). Nevertheless, social acceptance of the hospital death has become so widespread that clinical medicine now recognizes a new area of expertise, referred to as "palliative care" (Institute of Medicine 1997). The majority of us will most likely die in a hospital; deaths that occur elsewhere are frequently thought of as accidental and even preventable.

These realities of contemporary death in America bear two significant consequences relevant to this chapter. First, it can be rather difficult to die alone; yet, ironically, it can also be difficult to die in the company of our intimates. (This is why many hospitals and hospices have instated policies to enable close kin, at least, to be at the bedside throughout the day and night.) Second, death has become increasingly technocratic, and this is a marker more broadly speaking of late twentieth-century biomedicine in the United States. As a nation we are so aware of this trend that many of us now draft "living wills" (a peculiar oxymoron) in an attempt to dictate just how much care and technological intervention we wish to bear. While we fear the uncertainty and suffering of death, we have a horror of the prolonging effects of heroic efforts to protect life. We know all too well that ventilators, drug therapies, and other forms of what we call "life support" can sustain a body in stasis even when our intimates are certain we, as individual selves, no longer dwell "in there." The horror of the technocratic death is that it robs us of our free will or agency, and it dehumanizes us by transforming us rapidly into cyborg creatures, or human bodies devoid of personhood that are nevertheless dependent on machines for survival.

Thus, within the context of early twenty-first century medicine, we must ask, What comprises a good death (Das 1997; Harmon 1998; Nuland 1993; Seale 1998)? Does it hinge on quality medical care and the knowledge that heroic efforts were taken to keep us alive as long as possible? Is it based on the ability of kin to be there to bear witness to our last moments on earth? Or does it depend on our assumed ability to determine or even manage the outcome, so to speak, of our personal endings? These are deeply troubling thoughts that dog conversations among clinicians and ethicists, journalists, family members, and friends, as each struggles to reach some level of understanding about what to do with the dying and the dead.

The three central chapters of *Bodies, Commodities, and Biotechnologies* explore different aspects of technocratic biomedicine in the United States. To illustrate my points, I draw on the medical realm I refer to as "organ transfer." As I have long argued (Sharp 1994, 1999, 2001), among the more confounding aspects of the existing literature is the persistence with which professionals compartmentalize the domains of organ transplantation, procurement or retrieval, and donation. The most privileged of these three is transplantation, where we frequently celebrate the seemingly miraculous outcomes of sophisticated surgeries that, when successful, grant renewed life to formerly ailing (and often dying) patients. But a sad truth of transplantation is that it relies on suffering and loss in other quarters, either because it subjects robust and healthy people to dangerous surgeries as they willingly give up, say, a kidney to help another, or, even more extreme, because of the need to extract still vital organs from individuals who only minutes or hours before have been declared dead. I advocate, then, that organ transplantation be understood as inextricably intertwined with procurement and donation procedures. Thus, throughout this work I regularly employ the expression "organ transfer" to underscore this inseparability.

The ascent in this country of organ transplantation, as the outcome of organ transfer, is truly remarkable: within 50 years, surgeons have moved from highly experimental attempts to transfer body parts between dogs and, soon thereafter, identical twins to the point where over 24,000 transplants are performed annually in the United States. Regardless of the routinization of the procedure, transplants are still regularly hailed as medical miracles. When we think in precise terms about what is being accomplished, it does indeed seem miraculous. After all, organ transfer involves the retrieval of functioning body parts from sometimes living or, even odder still, from dead bodies, so that they may replace the ailing heart, lungs, liver, kidneys, pancreas, and intestine in ailing patients. Even more remarkable is the fact that these patients—or organ recipients, as they are called—survive the ordeal. Following their surgeries, they may live for many years supported by potent immunosuppressants and a host of other powerful drugs designed to counteract graft rejection. Organ transfer is thus simultaneously wondrous and strange. I argue that this is especially true when cadaveric donors are involved—that is, when organs are retrieved from the bodies of patients who have been declared dead.

I wish to stress that within this work my comments on organ transfer concern exclusively cadaveric organ donation and not the effects of cases involving living donors (for example, when someone gives one kidney to another, or a surgeon transects the liver of a living patient and gives one of the two parts to another family member in need). Furthermore, I speak only of legal practices based within the United States, and not of the clandestine global commerce in human body parts. Finally, unless stated otherwise, I refer specifically to transplantable, whole organs, not human tissues. This is because whole organs (and especially the heart) carry much greater symbolic weight than, say, skin, ligaments, and bone, so they inspire radically different social responses. As I will illustrate, very peculiar and, thus, deeply intriguing things happen, from an anthropological perspective, in this highly medicalized arena, where it is intrinsically understood by all involved parties that death can beget life.

I must stress, too, that in focusing on organ transfer in America, my purpose is not to demean this complex realm of medical practice. Rather than serving as social critic, I instead offer a set of critiques rendered possible by the anthropological stance of studying exotics at home (di Leonardo 1998). Ethnographic work is driven in large part by the understanding that a careful interrogation of mundane, and thus unquestioned, daily practices may in fact generate a wealth of cultural knowledge (Malinowski 1961 [1922]). Within this set of lectures, I am especially interested in the deeper meanings and paradoxes that are obscured by regular assertions of organ transplantation's miraculous qualities within my own society.

Bodies, Commodities, and Biotechnologies analyzes the intriguing connections between disparate bodies in medicine; I draw on ethnographic research in the realm of organ transfer in the United States as a means to explore these connections. This first chapter looks specifically at the dilemmas engendered by the medical fact that death can indeed beget life. As I will show, among the most perplexing consequences of organ transfer are questions of how professionals and the surviving kin of organ donors should mourn such an unusual category of the dead, when their valued parts have been dispersed to assist others in need. In chapter 2, "Body Commodities," I analyze the consequences of professional denial of body commodification, which gives rise to remarkably innovative (and, at times, subversive) responses from both organ recipients and the surviving kin of organ donors. Among the most intriguing developments are the ways that involved parties assert new ties of sociality, where the transplanted parts of deceased donors transform anonymous strangers into intimate kin. In the third and final chapter, "Human, Monkey, Machine," I explore the social ramifications of highly experimental attempts at body melding, where transgenic animals are imagined by scientific researchers as one day generating organs for human use.

Before we delve into the realm of organ transfer, I offer a few words about myself and what concerns me a scholar. A medical anthropologist by training, I have long been interested in questions of embodiment. As assumption that drives much of my work is that the body offers a rich, obvious, and even natural terrain on which to impose meaning (Douglas 1970; Mauss 1973 [1935]; Scheper-Hughes and Lock 1987). Questions that drive my work include: What meanings are assigned to the human body in various cultural contexts? When we decipher these, what do they tell us about how particular categories of people, or bodies, or body parts, are valued (or devalued) by members of that particular society? How do values differ according to various life stages? And what of death, and dead bodies? What larger structural elements of society—such as political and economic forms of power—might affect or play off of the meanings assigned to various bodies? The human body in many cultural contexts offers a symbolic framework for examining larger social questions. It is a crucial site for experiences ranging from expressive, artistic acts to profound forms of suffering. The body, then, is a significant cultural artifact.

I am first and foremost an ethnographer, and my work on the body, which now spans nearly two decades, has followed a circuitous route. My original (that is, dissertation) research began in a place quite remote from New York City, where I now live and work. In late 1986 I ventured to Madagascar, a large, remote island a few hundred miles off the coast of Mozambique in the Indian Ocean. For over a year there I worked with spirit mediums, women of a range of ages who, during trance states, temporarily embodied *tromba*, or the spirits of well-known dead royalty (Sharp 1993; 2000). Tromba possession is wonderfully theatrical because a possessed medium assumes a radically different style of dress, demeanor, and persona. When she occupies such a state, one quickly discerns that one is no longer in her presence but, rather, speaking to a spirit summoned from its royal tomb. Such behavior from mediums is considered perfectly normal. It is normal, too, to converse with spirits just as with any living human being, albeit with great respect, as when in the presence of royalty. Because mediums embody multiple selves, they frequently have a more complicated and, ultimately, more integrated sense of self than do the nonpossessed. This is in part because, within Malagasy culture, self-reflection is an oddity. In everyday discourse, one does not draw attention to the self, and the grammatical structure of Malagasy as a language often dictates use of the passive, rather than the active, voice. Frequently, too, the first-person pronoun is not uttered, only implied. Just as one does not draw attention to the self, one does not think of oneself. Mediums, however, frequently have complex personal histories fraught with suffering, and, as I have argued elsewhere, their spirits' biographies can become entangled with, complement, or deepen their understandings of their private predicaments (Sharp 1995). (This idea of embodying multiple yet integrated selves will prove helpful to my discussions that follow on organ transfer.)

Spirit mediums frequently work as healers, drawing their power from the sacred knowledge of these royal ancestors, and they may engage in a brisk trade several days out of most months, assisting clients who make appointments to consult a specific spirit. In my desire to learn of the significance of royal power in the region, I was frequently instructed by mediums and others—from peasant clients to living royalty and their retainers to schoolteachers and government officials—to seek answers to specific questions from the *tromba* spirits. In response, I made appointments so that I could interview mediums and, subsequently, their individual spirits. In essence, I spent much of my time over the course of a year interviewing the ancestral dead.

The audiences that gathered during the occasion of these Schoff Lectures at Columbia University, in one of the most cosmopolitan cities in the world, might well envision such practices as exotic and strange, as might readers. From the perspective of my friends and associates in Madagascar, however, organ transfer is beyond exotic: the thought of it is horrific. Malagasy are well known within anthropological circles for their elaborate mortuary practices, so much so that it can be claimed that, at least in contrast to dominant North American contexts, Malagasy spend their lifetimes preparing for death and entombment. Human bodies are dangerous things in Madagascar: in their decaying state, they embody filth. But bodies are also handled lovingly, and in some regions of the island they are tended for many years, removed periodically from their tombs and wrapped in new shrouds of hand-woven cotton or silk. In this process, they are remembered discretely as individual ancestors; only once a person's name is forgotten are their pulverized remains mixed with those of others. There are elaborate rituals, too, surrounding the care of rulers' corpses, whose effluvia are collected and discarded by special retainers who secret them away in sacred forests. Parts that remain—fragments of bone, nails, hair, teeth—are transformed into coveted objects that embody the essence of royal power (Bloch 1971; Feeley-Harnik 1984; Sharp 2000).

Organ transfer is unthinkable to Malagasy because it defies cardinal rules surrounding body integrity. Body parts that are shed or removed—ranging from a circumcised boy's foreskin to a mother's placenta to the flesh of a deceased ruler—are never simply discarded but, rather, are ingested or buried or hidden where they can be protected by kin or other trustworthy caretakers. And only royal corpses are deliberately cleaned of their fleshy parts, the harder, durable remains preserved with care for ritual purposes by the dynasty's successors. It is only witches who, prowling the night, take body parts and blood from human beings for selfish and nefarious purposes. A truly horrific image to Malagasy, then, is that of a surgeon removing the heart of a dead person in order to place it in the body of a living patient. Put plainly, organ transplantation is not a very popular idea in Madagascar (cf. White 2000).

As should be clear by now, anthropology is a comparative science. Although Madagascar is radically different in many obvious ways from the culture of technocratic biomedicine in America, these sorts of observations have proved helpful to my research on organ transfer. As I will illustrate throughout this volume, the human body and its transplantable parts reveal much about the values we assign to the private self, sociality and intimacy, humanity, and human nature. At this point, I turn to the clinical management of an unusual category of the dead.

MANAGING THE CLINICALLY DEAD IN AMERICA

Cadaveric organ donation currently accounts for approximately half of the annual total of about 13,000 organ donors in this country.[1] Although living donation is on the rise (and has in fact surpassed cadaveric donation, at least for kidneys), the dead donor is nevertheless an inescapable necessity of survival for many organ transplant recipients. Most lungs, livers, and hearts, for instance, are procured from the dead.[2] Within American society, organ transfer is shaped by a set of dominant ideological principles that promote the process of cadaveric donation as a great social good. I will highlight briefly those principles most central to the arguments presented throughout this book. (For more details see Sharp in press.)

First, it is illegal for whole organs to be bought or sold in the United States. Rather, organ transfer's success relies on individuals who are, quite literally, willing to give of themselves to others in need. Regardless of the fact that organ transplantation is among the most profitable medical specialties in this country, human organs are always "donated," and they have long been described as "gifts of life" (a phrase derived from the blood industry, and currently applied as well to gestational surrogacy).

Second, transplantable human parts are understood as precious things, rendered all the more so by a growing anxiety over organ scarcity, because today the demand for organs far outpaces the supply. As of early November 2005, slightly over 90,000 people were waiting for transplants, while 27,037 actually received organs over the course of a full year in 2004.[3] This notion of scarcity informs very specialized ways of talking about the human body and its parts. Within some professional circles, for example, organs are even described as one of our most precious national resources.

Third, because so many transplants depend on cadaveric donation, the legitimacy of organ transfer in legal, medical, and social terms necessitates that all involved parties embrace the legitimacy of brain death criteria (Lock 2002). This is because head injury—resulting from a traffic accident, a gunshot wound, or a massive stroke, for instance—remains the primary cause of death for organ donors in this country. (Other categories of death are increasingly leading to organ donation, but I will not go into this here; see Sharp in press.)

It is important to understand that organ transfer involves a range of professional and lay parties, all of whom must embrace certain medical understandings of death for the process to succeed. Participants include transplant surgeons, nurses, and social workers; professionals who work for organ procurement organizations (OPOs); organ transplant recipients; and the surviving kin of organ donors. As members of all four groups regularly attest, among the more troubling aspects of brain death criteria is that seeing is not necessarily believing. This is because ventilators and other related technologies (which we regularly refer to in other contexts as forms of "life support") enable a brain-dead body to breathe artificially. This then stimulates the heart to pump oxygenated blood throughout the body, which in turn helps other organs to remain vital (or, as some would say, "alive"). As a result, the brain dead retain their color, are warm to the touch, and appear to breathe on their own (albeit to the rhythm of the ventilator). As any ICU staff member or procurement specialist will explain, though, this is a highly precarious clinical state that requires careful monitoring if organs are to be acquired before the body "crashes." Brain death is thus a most peculiar state of nonbeing: we should not think of such patients (if we can call them that) simply as *unconscious* but, rather, more appropriately, as terminally *devoid of consciousness*. As many interviewees have expressed to me over the years, the brain dead may appear as though they are asleep, but in reality they are no longer "there."

The success of organ transfer in this country relies therefore on the acceptance that brain-dead organ donors are devoid of personhood. This is driven by a widespread (though far from unanimous) biocultural understanding that the mind (or brain) is what renders us unique as human beings and as known, social individuals (Damasio 1999). This means that when our brains permanently cease to function, we have lost our humanity forever. Once a patient is declared brain dead, he or she is rapidly transformed into what various authors have dubbed the "living cadaver," a body maintained artificially by a host of machines in anticipation of the surgical removal of its organs (Hogle 1995; Lock 2002).

Framed in this way, the necessary work of organ procurement is extraordinarily difficult. Transplant surgeons—like all physicians—most certainly suffer deeply from the losses of their patients; their work is nevertheless understood within this culture as a life-giving and even miraculous mission. In contrast, procurement professionals must confront death every day. As the employees of nonprofit organizations that specialize in the art of organ retrieval, they often help identify and sometimes assist in diagnosing brain death in patients; help track down next of kin; and then must approach these kin to explain brain death criteria and elicit support for organ donation. Although signed organ donor cards assist these professionals in their work, they nevertheless may have to endure insults and rage from family members deeply immersed in early stages of shock and mourning. Procurement professionals also regularly provide emotional support to the bereaved not simply during the period of hospital care but for weeks, months, and even years beyond donors' deaths (Sharp 2001).

Yet another related aspect of the ideology that drives organ transfer is that this medical realm is rife with taboos, evident in what historian Ruth Richardson refers to as forms of "semantic massage" (Richardson 1996). As noted earlier, it is illegal to buy and sell human organs, and thus in the United States they are referred to exclusively as "gifts," not "goods." Furthermore, a range of labels are applied to human beings whose bodies are sources for transplantable organs. To ICU staff, they remain "patients" up until and even throughout procurement surgery. Procurement staff, however, rarely speak of "patients" (especially among themselves), preferring instead (and sometimes even insisting on) the use of the term "donors" as a means to underscore that they cannot be patients because they are dead. The transplant literature authored by surgeons, though, contains such terms as "cadaver" and "neomort." This terminology deeply offends both procurement staff and, of course, the surviving kin of donors. Kin also abhor the use of the term "donor," especially prior to procurement surgery. Instead, they almost exclusively employ personal names and other intimate forms of address. Referring to a loved one as "the donor" or even as "the patient" is perceived as a disturbing form of dehumanization.

THE UNNAMED DEAD

In addition to the regular reworking of terminology assigned to donor bodies, one also encounters practices that cloud or altogether erase donors' identities. A final dominant ideological premise I wish to address is that transplantation's success hinges on donor anonymity. Two factors are involved. First, it is considered psychologically dangerous for organ recipients to identify emotionally with their donors. Recipients are told regularly by their surgeons, for example, that their hearts are "mere pumps" or that their kidneys or livers are "filters." Some recipients embrace this so willingly that they may draw on such imagery to the point that they liken their surgeries to elaborate forms of mechanical or, more precisely, automobile repair. Recipients also learn very quickly to speak only privately—or, at the very least, out of earshot of transplant professionals—about any sentiments they feel for their donors. Worse yet is the idea that they are somehow transformed because their bodies now house, for example, a younger heart, or a part derived from someone of another gender. The information supplied to organ recipients about their donors is limited to age, gender, and perhaps the region of the country where they lived and, maybe, on occasion, their family status. Often they know little or nothing about the cause of a donor's death, especially if it resulted from a homicide or suicide. Personal names are always withheld. Thus, recipients are left to imagine who their donors might have been, and anyone who assigns meaning to their organ's origins may well be shuttled down the hall for a psychiatric evaluation, only to emerge with the diagnosis of "Frankenstein Syndrome" (Beidel 1987)—that they suffer from the delusion that they have been pieced together with parts derived from more than one body (even though, in fact, they have).

Anonymity is also guarded out of respect for donor kin, for procurement specialists have long asserted that privacy matters. The logic is that donor kin have already suffered enough, and inquisitive, though well-meaning, recipients could undermine the healing process that should follow a tragic loss. Interestingly, very particular messages are employed to promote donation among donor kin, and they conflict in profound ways with those that circulate on transplant wards. Among the most frequently used promotional messages is that through donation the lost loved one can "live on" in others, and organ donors are indeed regularly understood by their surviving kin as doing precisely this. The fact that many surviving kin are parents who have lost their children plays an important role in shaping such sentiments. But this idea that the donor lives on, at least in part, in others is a far cry from the characterization in which the organ recipient's body is likened to a repaired automobile. As I will illustrate in my second chapter, this profound form of ideological disjunction figures prominently in shaping attempts by recipients and donor kin to establish ties of social intimacy with one another.

THE IMPOSSIBILITY OF RECIPROCITY

Marcel Mauss, the nephew and pupil of Emile Durkheim, argued approximately eighty years ago that gifts "are in theory voluntary, disinterested and spontaneous, but are in fact obligatory and interested." Gifts (*prestations*) must be repaid, and it is the act of reciprocation that ultimately lies at the heart of social relationships (Mauss 1967:1). In this light, organ donation becomes highly problematic: as an anonymous form of gift-giving, the process renders it impossible for recipients to compensate the donor or his or her surviving kin for what they have done. Given that organs, as precious gifts, are derived from the bodies of the dead and, furthermore, extend and save the lives of others, the desire to offer thanks is indeed a heavy burden to bear. Even if one could thank surviving kin, how could one ever hope to offer something of similar value in return? The fact that long-term use of immunosuppressant drugs excludes transplant recipients as future organ donors further complicates the picture. This horrible paradox—the inability of recipients to repay such profound acts of kindness—is what sociologists Renée Fox and Judith Swazey have referred to as the "tyranny of the gift" (Fox and Swazey 1992). How, then, do we give thanks for so extraordinary a gift from the dead? One way is to memorialize them.

REMEMBERING THE DEAD IN MEDICAL CONTEXTS

Commemorating the deaths of those who give of themselves to medical science is a growing trend in the United States. Medical students across the country, for instance, regularly participate in end-of-the-year ceremonies staged in honor of the cadavers that enabled them to study human anatomy. In some instances, the kin of those who "donated their bodies to science" may be invited to attend, so medical students can mingle with them. Such circumstances enable students, if they wish, to strike up conversations with family members about the specific people whom they themselves have dissected (Dixon 1999; Roach 2003).

Most readers will find such practices macabre. What I propose, however, is that we step back a bit and try to think carefully about this need or longing to commemorate the deaths of such strangers. As I will show, any desire to memorialize this unusual category of the dead must overcome several obstacles. Specifically, in the realm of organ transfer, cadaveric organs are given anonymously, and thus their origins remain obscure. This is further exacerbated by the fact that these organs have been severed from the bodies that originally housed them and dispersed throughout a region or across the nation. An inescapable characteristic of cadaveric donation in the United States, then, is that it violates the integrity of individual bodies: first, because the donor body is disassembled, so to speak, and second, because organ recipients in turn must experience radical forms of body repair.

Even more troubling is the fact that the ideology driving organ transfer in this country regularly celebrates life while denying death's presence. This is evident, for instance, in the expression "gift of life," and the currently popular slogan, "Donate Life," as featured on innumerable print materials and Web sites nationwide.[4] Ten years ago the majority of the nation's OPOs bore names that reflected the location and nature of their work, as with IOPO, or Indiana Organ Procurement Organization, for example. Today, however, corporate names more frequently deny death's presence. Consider, for instance, Gift of Life of Delaware and Michigan; LifeCenter in Idaho, Alabama, and Washington; LifeBanc, Life Connection, and Lifeline in Ohio; Life Choice in Connecticut; LifeNet of Virginia; LifeShare in North Carolina; LifeSource of Minnesota; and Florida's Life Link, Life Alliance, TransLife, and LifeQuest. Organ recipients have developed their own specialized way of celebrating life over death, for many speak of their transplants as "rebirths," and they may even mark its date with a party where they and loved ones ingest a cake baked in the shape of the transplanted organ. Organ transfer, clearly, has generated a remarkable form of medicalized communion, not to mention symbolic autocannibalism.

A range of metaphorical constructions similarly obscures the origins of procured organs. A genre that dominates professional language and promotional materials involves what I refer to as the "greening" of the donor body (Sharp 2001). After all, we refer to the surgical replacement of organs as "transplantation," and surgeons regularly speak of "harvesting" organs from donors. This green imagery figures prominently, too, in a lively array of promotional materials that draws on plant imagery while obscuring the presence of the dead body. For instance, in materials that describe brain death, trees standing at the edge of a field might evoke a sense of loss or emptiness, or images of flower gardens, bouquets, and potted plants serve to underscore rejuvenation, fertility, and rebirth. As I have described elsewhere, among the more elaborate renderings to date is a colorful poster published by the National Association for Transplant Coordinators (NATCO) some years back, intended as a means to thank the nation's array of organ donors. The iconography of this image is quite complex: the dominant portion of the painting consists of a young man watering a flower garden in a world where water drops are reminiscent of tears, and where clouds assume the shapes of pancreases and livers (see Sharp 2001:122). Greens imagery is so pervasive in the realm of organ transfer that even children regularly rely on images of flowers and other plant life as a means to promote organ donation when asked by hospital staff or organizers of donation campaigns to design posters and other promotional materials. Another common motif is the butterfly, an image employed for some time by the National Donor Family Council (NDFC), a group that advocates for the rights of donor kin. Butterflies, like vegetation, offer "natural" symbols, so to speak, of regeneration and rebirth. The fact that organ recipients regularly speak of their surgeries as rebirths further assists in naturalizing such imagery.[5]

What, then, of memorial forms? A now well-established practice in this country involves the planting of donor gardens. These range from modest projects consisting of one or a handful of trees placed on hospital grounds to a full grove of saplings, or one tree planted in, say, a community park in honor of the donors from that region who gave of themselves. Some gardens now also display a newly developed hybrid known as the Gift of Life Rose. Other memorials are elaborately landscaped, and they may include benches, memorial plaques, and large, engraved stones that together offer a contemplative space where visitors might pause and rest. Donor gardens are now so plentiful that I frequently ask not whether a hospital or OPO has one but where it is located.

Clearly, I am deeply intrigued by the imagery that abounds here, and I have written on this at length (Sharp in press). What fascinates me the most is how such memorial projects stand out as bureaucratized forms of mourning. A dominant practice nationwide involves annual tree-planting ceremonies as a means to honor the dead while celebrating renewed life through transplantation (see figure 1.1). A particularly striking aspect of these events is that they strive to bridge the divide—albeit temporarily and in carefully orchestrated ways—between organ recipients and donor kin. Without such mutual involvement they fail as legitimate, public events. The tone, however, is far from celebratory. Recipients and donor kin are regularly invited (and even expected) to place some earth on the tree's roots, not unlike in a cemetery burial, when the survivor shovels a bit of soil into the loved one's grave. Recall, too, that these ceremonies are staged (and for the most part funded) by transplant units and hospitals, or by individual OPOs and tissue banks. Therefore, they simultaneously honor donors and celebrate professional successes. Thus, whereas the rhetoric surrounding such stagings emphasizes the importance of com-

FIGURE 1.1. An annual tree planting ceremony, circa early 1990s. *Photo by the author.*

munity, giving thanks, and even reciprocal participation, they should also be considered first and foremost as public relations events. As transplantation has grown as a medical practice nationwide, these stagings have become more frequent and more elaborate over the years, especially during the 1990s.

Memorials have also grown in size. The most extravagant to date is the National Donor Memorial, a project that dominates the grounds of the new corporate headquarters of UNOS in Richmond, Virginia (see figure 1.2). The memorial encompasses 10,000 square feet of outdoor space and bears a price tag of approximately $1.2 million. It draws on motifs already well established in other locales: there is, for instance, a memorial grove of sapling holly trees as well as a butterfly garden. Most intriguing, though, is the "Wall of Tears" that marks the entrance to the memorial, and the "Room of Remembrance,"

FIGURE 1.2. The National Donor Memorial, located on the grounds of the United Network for Organ Sharing (UNOS) in Richmond, VA. UNOS is a nonprofit organization under the direction of the U.S. Department of Health and Human Services; it oversees the procurement and allocation of transplantable human organs. *Photo by the author, spring 2004.*

FIGURE 1.3. The wall of names, the dominant structure within the Room of Remembrance of the National Donor Memorial, Richmond, VA. *Photo by the author, spring 2004.*

the central of three outdoor spaces. This room features a wall that displays the words "Hope—Renewal—Transformation" and the first names of a range of donors culled from the UNOS database (see figure 1.3). Names such as Etsuko, Melissa, Enrique, and Brianna are displayed on the memorial wall, and they have been selected specifically in order to emphasize, in the words of one employee, the "diverse origins" of the nation's organ donors. As first names only, they stand as a generic category, however exotic they may seem to some visitors.

What, then, are we to make of this pervasive use of greens and other memorial imagery? Tree planting in and of itself serves as a common form of commemoration in a host of contexts. Across the United States, a range of gardens honor financial sponsors (an altogether different category of donor) or the victims of crimes or incurable diseases, or they serve as ecological projects designed to stave off urban or suburban sprawl. All, in their own ways, celebrate the theme of renewal. When organ donor gardens are considered in their entirety, however, the iconography of plant life, set against the cenotaph and stone marker, reveals an altogether different set of ideological goals. As I will show, the symbolism of these specialized, and medicalized, projects is highly reminiscent of war memorials, or projects that honor an altogether different category of the dead.

HONORING THE NATION'S DEAD

A well-established tradition in this country, employed to commemorate the lives of fallen soldiers, involves the planting of groves of trees in public space alongside stone markers, statues, and the like. One need only venture to Central Park in New York City to encounter one such example. Any visitor strolling down Fifth Avenue might enter the park around East 69th Street and walk west. A minute or so from the park's boundary stands a grove of over a dozen pin and red oak trees (see figure 1.4). As a commemorative stone explains, these trees are dedicated "To The Dead of the 307th Infantry A.E.F. / 590 Officers and Men / 1917–1919." Another plaque, found at the base of a tree, labels it specifically as a "Memorial Tree" where "They Sleep." Scattered throughout the grove are 13 other commemorative plaques, each similarly resting at the foot of an individual tree and bearing a set of soldiers' full names, from as few as a dozen to more than 60.[6] A complete assemblage of these names also can be found on the obverse of another cen-

FIGURE 1.4. Memorial grove erected in honor of the 307th Infantry, Central Park, New York City. *Photo by the author, spring 2004.*

tral stone marker. To the side of the grove stands yet another, erected more recently and placed to the side, for soldiers who had belonged to various lodges of the Knights of Pythias. This repetitious quality of naming exposes an unending anxiety that, with time, the names of these soldiers, from what is sometimes referred to as the "Lost Battalion," may indeed be lost and thus forgotten.

Whereas this memorial grove in Central Park illustrates a long-standing practice in this country of planting trees in honor of the dead, another national project has most certainly inspired the architects of the National Donor Memorial, too.

This is the Vietnam Veterans Memorial in Washington, DC, which lies only 100 miles away from Richmond (see figure 1.5). The power of the Vietnam memorial lies in its insistence on naming the dead, in the form of a wall upon which over 58,000 names are engraved (Sturken 1997:58–63). Other, more modest projects across the nation draw on such iconography, too, as in a far more modest memorial space erected recently near the center of Ripon, Wisconsin, a town where I attended high school (see figure 1.6).

Interestingly, the National Donor Memorial visually bears much in common with these war memorials, where the centerpiece of the site is a stone wall upon which names have been engraved. But unlike the Vietnam memorial (sometimes referred to simply as "The Wall") or the Ripon site, the National Donor Memorial fails to commemorate the very individuals it seeks to honor precisely because of the reluctance or even distaste for including organ donors' complete names. Financial and other pragmatic constraints may play some role, for one might well argue that UNOS should not shoulder the burden of adding new names each year as the nation's "honor roll" of donors increases. (Of course, in other settings where donors are philanthropists, host institutions may not hesitate to leave room for future names.) UNOS must also be concerned that such an endeavor would require obtaining consent from donor kin before listing the names of those initially understood as anonymous.

As such, the National Donor Memorial is emblematic of the problems that plague donor garden projects throughout the United States (and, potentially, medical memorials more generally). Conceived in part as a public relations tool (or, in the words of UNOS employees, an educational project that schoolchildren can tour to learn about organ donation), it evades more troublesome aspects of organ transfer in America. Why must the dead remain unnamed? Is anonymity so important that we must trun-

FIGURE 1.5. The Vietnam Veterans Memorial, Washington, DC. *Photo by the author, winter 2005.*

FIGURE 1.6. Plaque and inlaid, etched stones from the Ripon Area Veteran's Memorial Walk of Honor, Ripon, WI. *Photo by the author, summer 2005.*

cate identities by eliminating references to family origins? How is one to read the names of the dead when they are displayed simultaneously as an exotic assortment of unusual first names that, when viewed together, nevertheless merge into a generic category of unknown souls? Who were these people? How and when did they die? And what of the sense of sorrow and individual loss, when surviving donor kin visit this memorial site? As is made all too clear in a recent UNOS publication, donor kin do in fact long to encounter the personalized names of those whom they mourn. Just as visitors to the Vietnam memorial regularly kneel before it to make rubbings of engraved names, donor kin might pause to touch names familiar to them. Among the most poignant images from the dedication of the UNOS National Donor Memorial in fact shows Norma Garcia placing her hand on "Jasmine," the name of her own daughter who, at thirteen, became an organ donor (UNOS 2003:5, 8–9).

SUBVERSIVE MEMORIAL PROJECTS

Given these parallels—and discrepancies—we must ask how lay participants perceive such memorials. As I have discovered through interviews, organ recipients are regularly and deeply moved by these projects, and often speak of the catharsis that accompanies their participation in memorial events or during visits (which may approximate pilgrimages) to these sacred sites. Interestingly, in contrast, donor kin more often express reticence and, at times, even disgust for such memorials. On several occasions donor kin have uttered quietly to me how distasteful they find tree-decorating or planting ceremonies, likening them to erecting Christmas trees solely to assuage recipients' guilt, or even, in the words of one father, describing the memorials as "mass graves." Donor kin underscore the rage they feel at being silenced by projects that simultaneously consign individual donors to a generic category of dead, while also denying them, as the kin who remember these individuals, the opportunity to tell their stories and speak of personal pain and loss.

In response, in recent years, some donor kin have assumed an activist stance. An especially effective group is the National Donor Family Council (NDFC), whose emblem is a pink butterfly. From its inception, the NDFC has challenged the dominant assumption that donors' identities must remain obscure. Among the most prominent artifacts of this movement is the Patches of Love Quilt Project (see figure 1.7). Today this national quilt consists of approximately 1,500 squares submitted from diverse regions of the country by the kin, friends, and other loved ones of organ donors. Clearly modeled after the AIDS Quilt (see figure 1.8), the Patches of Love Quilt has played a significant role in

FIGURE 1.7. Sample squares from a panel of the Patches of Love quilt project of the National Donor Family Council, an organization housed within the headquarters of the National Kidney Foundation in New York City. *Photo by the author, mid-1990s; published with permission from the National Kidney Foundation.*

subverting the ideological assumption that organ donors should remain anonymous and, thus, unnamed.

As with the AIDS Quilt, involved NDFC staff impose no restrictions on the content of their quilt's squares, and, as a result, submitted pieces regularly display donors' full names, and often their dates of birth and death. The latter is especially important because it is often the same date as a recipient's transplant surgery. Yet other parallels can be drawn with the Vietnam Veterans Memorial, for recipients who view the donor quilt can conceivably search for their donors' names. The quilt is also embla-

FIGURE 1.8. Sample squares from the AIDS Quilt of the Names Project, on display in Washington, DC in October 1996. *Photo by the author.*

zoned with donors' photos, and many squares are crowded with personal mementos. As is true of the other memorial projects, the Patches of Love Quilt is simultaneously a visual and tactile experience, with individual squares displaying everything from athletic letters to state pins, pieces of clothing, costume jewelry, tiny tools, and toys. Some squares also make direct or oblique reference to how a donor died: a hunting accident, a motorcycle crash, a murder. Because NDFC requests that submissions be accompanied by personal statements, quilt squares are carefully catalogued by the organization's archivist, who is herself a donor mother. And just as the AIDS Quilt tours the nation, so do panels of the Patches of Love Quilt. It always appears at the biannual Transplant Olympics, and OPOs, transplant units, and UNOS regularly ask to display it in their lobbies, wards, and corporate offices. Quilting has become a national pastime in the realm of organ transfer in America: several OPOs now have their own quilting projects, sometimes to honor the growing number of donors handled by their staff; in other instances, a highly personalized quilt, sewn for a particular donor, stands in for the whole. In this sense, donor kin and OPO staff, too, on occasion, work to piece back together, so to speak, the bodies of anonymous donors whose organs have been dispersed to save the lives of patients in need.

LOSS AND LONGING

When we consider such memorials in comparative terms, an especially striking relationship emerges between military and medical projects. In times of war, we are plagued by two related anxieties: that the inevitable casualties will include those we love, and that we will be unable to retrieve their bodies for proper burial (Webster 1996). For instance, an article in the October 25, 2004 issue of *The New Yorker*, written by Caroline Alexander and entitled "Across the River Styx," documents this desire as driving the unending search for the bodies of soldiers still missing in action in Vietnam (Alexander 2004). A requisite in the aftermath of war, to erect memorials to those killed in action, has characterized this nation's responses to a range of wars and battles. Small towns and villages throughout the northeastern region of the United States are smattered with over a century's worth of monuments erected in honor of the fallen dead, ranging from the Revolutionary War and Civil War to the War of 1812, the two World Wars, and the conflicts in Korea and Vietnam. As noted above, such projects insist on providing the full names of the dead, frequently accompanied by the villages and towns whence they came, particularly in small-town settings. These projects ensure that the names of the dead become—and will remain—part of the public record.

Interestingly, we do the same with large-scale tragedies during peacetime. Heated current debates surrounding how best to memorialize the World Trade Center disaster in New York City illustrate how prominently such projects figure in the process of national mourning. Other, completed projects offer profound insights into how wartime iconography bleeds, so to speak, into other contexts of life. Consider, for example, the Oklahoma City National Memorial, dominated by an outdoor field of 168 chairs, each engraved with the name of one of the victims of the bombing of the Alfred P. Murrah Federal Building by Timothy McVeigh on April 19, 1995. Another is the Lockerbie Memorial Cairn—located, notably, in Arlington National Cemetery (interestingly, a predominantly military cemetery) in Washington, DC (see figure 1.9). Erected in the aftermath of the 1988 plane crash outside Lockerbie, the memorial consists of 270 blocks of reddish sandstone quarried from this region of Scotland, one block for each person (259 passengers and 11 local residents) killed when a bomb exploded on board Pan Am flight 103. In addition, engraved around the base of the memorial are the full names of the 270 dead.

Set against such examples, the inability to name the dead emerges as a mark of national failure. As this book goes to press, yet another national tragedy that began in late August 2005 further underscores my point: in the wake of the devastation of Hurricane Katrina, Americans witnessed the blatantly callous disregard for the dying and the dead in New Orleans within several branches and agencies of the federal government. Among the more horrific images to emerge from this national tragedy were photographs and stories of the regular abandonment of unnamed, dead bodies.

FIGURE 1.9. The Lockerbie Memorial Cairn, Arlington National Cemetery, Arlington, VA. The cairn commemorates the 2,070 people (who were either passengers or on the ground) who died in 1988 as a result of the terrorist bombing of Pan Am flight 103 near Lockerbie, Scotland. *Photo by the author, spring 2004.*

In contrast, the intertwined sentiments of loss and shame regularly shadow forensic work at the war front, where soldiers' bodies, in entirety or in part, are retrieved with meticulous care. Especially disturbing to Americans (and people of many other cultures) is the thought that soldiers' bodies have been left (that is, abandoned) on enemy soil. The suffering this engenders may have no end. For this reason, the Tomb of the Unknowns has long stood as a potent national symbol of sorrow and loss. It is, in fact, among the most popular sites visited by tourists in Arlington National Cemetery today (see figure 1.10).

Such anxieties and forms of collective suffering are hardly confined to our North American context: various European nations have erected monuments to those who were abandoned on the Russian front, and to colonial conscripts who died far from home during the two World Wars (Webster 1996). In Fréjus, on the Côte d'Azur in southern France, for instance, there are several memorials specifically honoring colonial subjects who were conscripted to serve in military campaigns in Europe and Indochina (see figures 1.11 and 1.12).[7] People as far away as Madagascar insist on the necessity of honoring—and remembering—dead conscripted soldiers, and formidable anxieties focus squarely on those who perished far from home. Within Madagascar, the spirits of dead soldiers cast overboard at sea can work their way home and wreak havoc on the living, haunting waterways, bridges, and the dreams of their progeny, whom they are certain have forgotten them (Sharp 2002:ch. 5).

Memorials erected to the unnamed dead, then, are troublesome things. Sometimes they are merely callous acts of convenience. An overly bureaucratized response insists that dead bodies breed too much sickness and thus more death, so they must be disposed of rapidly and efficiently. When mass graves are employed, however, they most often are reserved for the enemy. To lump bodies together anonymously is to debase their value by denying their individual humanity. As such, the mass grave is readily understood as a deliberate attempt to defile the enemy body. When the purpose is also to conceal

FIGURE 1.10. Tomb of the Unknowns, Arlington National Cemetery, Arlington, VA. Among the most popular sites visited by tourists each year within the most famous military cemetery in the United States, it honors unidentified dead who have died during conflict. *Photo by the author, spring 2004.*

FIGURE 1.11. Mémorial des Guerres en Indochine (Memorial to the Wars in Indochina), Fréjus, France. *Photo by the author, summer 2003.*

FIGURE 1.12. Mémorial de l'Armée Noire (Memorial to Black Soldiers), Fréjus, France. *Photo by the author, summer 2003.*

massive numbers of deaths, this form of burial denies surviving kin their right to grieve (Verdery 1999).

What, then, of medical memorials? Their intent is hardly to debase the dead or to hide them away. When compared to military projects, however, we come to realize that rarely are such memorials simply about giving thanks or marking history. Instead, they offer powerful forms of redemption, in which the act of naming figures prominently as a means of catharsis for those in mourning. In this light, there are clearly two faces to organ donor anonymity. Although officially, memorial projects serve to protect the privacy of involved lay parties (that is, organ recipients and surviving donor kin), they also relegate individual donors to a generic category of the dead. When the dead remain unnamed, surviving kin are forced to mourn their losses privately and even, it seems, in secret.

As I have sought to show within this chapter, medicalized deaths are highly problematic for a host of reasons. First, routine technological interventions intended to protect and save human lives rapidly dehumanize us as patients. Second, when we are under the care of medical professionals, we risk dying alone, among strangers rather than our intimates. And, finally, when understood as a special form of medical death, organ transfer complicates our expectations for what ultimately defines a "good death" in America. The goodness of organ retrieval rests with its ability to prolong or save the lives of others, but it is also a troublesome affair because it requires the partitioning of the body. The dispersal of organs to hospitals throughout the country necessitates destroying an individual body's integrity. This selfless act of generosity grants new life—or, as recipients often say, a rebirth. The ideology that drives organ transfer thus denies that the donor-as-person remains embodied even in a brain-dead state. Against this, memorial projects intended to commemorate the selfless act of giving unintentionally also deny the legitimacy of mourning the beloved in public as a cherished individual. The insistence that the dead remain anonymous is too great a burden to bear for many donor kin. This is why they have increasingly banded together to generate memorial forms that celebrate individual donors' lives beyond the grave. Just as the AIDS Quilt—sponsored, interestingly, by the "Names Project"—insists that forgetting is a dangerous social act, the Patches of Love Project likewise enables the unrestricted and open display of grief.

As I will explain in chapter 2, yet another intriguing response involves creative strategies employed by donor kin and organ recipients who together strive, in a sense, to reconstitute the dead bodies of organ donors. Defying the taboos long imposed by transplant and, to a lesser extent, procurement professionals, members of these lay parties are now, in ever greater numbers, seeking out one another to establish new bonds of intimacy. Although the anonymous dead are relegated to a generic category of the unknown by transplant professionals, the insistence in other quarters that donors have names means that they also have personal histories, histories of profound value to their surviving kin and to the recipients whose bodies now house their transplanted parts. As they seek out and find one another, donor kin and organ recipients stake out new territory in the realm of organ transfer, where bonds of intimacy are forged as seemingly natural responses to the highly unnatural medical practices so central to transplantation's success in America.

TWO

BODY COMMODITIES

The Medical Value of the Human Body and Its Parts

MARKETING THE HUMAN BODY

"A commodity appears at first sight an extremely obvious, trivial thing. But its analysis brings out that it is a very strange thing," wrote Karl Marx in *Capital*, volume 1 (Marx 1967:163). Within the United States, one regularly encounters a propensity to commodify things in peculiar ways, and this is particularly true of the human body and its parts. By 1992, the marketability of the human body had become so pronounced in this country that it prompted Jim Hogshire to author a guide on how to "sell yourself to science" (Hogshire 1992). Viable options included participating as a research subject in Phase One drug trials and reaping profits from such "liquid" and other assets as blood, urine, milk, eggs, sperm, and hair. At the time, Hogshire estimated that a human body, if dead for less than 15 hours, was worth $50,000 on the open market, whereas a living subject, through "guinea pigging," might earn around $100 per day, plus room and board.

Others, such as lawyer and policy analyst Andrew Kimbrell, have traced with meticulous care the expansion of what he called the "human body shop" in late twentieth-century medicine (Kimbrell 1993). Lori Andrews and the late Dorothy Nelkin, combining their expertise in law and medical sociology, have addressed thorny questions about body "ownership" in contexts where harvested organs, fetal tissue, and reproductive and genetic biomaterials are highly marketable commodities (Andrews and Nelkin 1998). But the use of the human body and its parts for lucrative gain is hardly new to the West, cautions historian Ruth Richardson (Richardson 1987, 1996). Throughout Europe, for instance, human bodies have long been sources for sacred, venerated objects; and the corpses of criminals, paupers, and others generated handsome incomes for those willing to raid the scaffold or gravesite to answer demands from dissectors, anatomists, and surgeons. Chapters of American history are also marked by pronounced forms of body commodification, ranging from the enslavement of living human beings to the now common marketing of body products associated with reproduction, as well as other tissues reaped from the dead.

A prominent debate within the contemporary realms of ethics and the law concerns the biomedical use of human bodies. Discussions focus regularly on themes of individual ownership as well as more widespread notions of community stewardship. The former concern—that is, our individual rights to our own bodies—emerges as inherently American. In contrast, in nations with a strong socialist bent, such as Sweden, for example, attempts to assert individual rights to corpses seem absurd, for as the state cares for its citizens when alive, the citizen's body parts are offered willingly to others in need at the time of death. (Some experts argue that this accounts in part for why Sweden's rate of donation answers its current demand for transplantable organs.)

In American discourse on the body, however, we regularly pose questions that attest to our sense that our bodies belong to us. If we can will our bodies to science, can we—or can our heirs—then claim that we own our bodies, or at the very least, assert rights to how they will be used? Does this mean that our bodies are our own property? If reusable parts can be extracted from our bodies, who holds rightful claims to these? Can or should we hold sway over the body's parts if they are of value on the open market? Should we be able to exert direct control over how these might be bought, sold, or used? And what are the dominant social sentiments regarding the willingness to be transformed into—or reduced to—a wide assortment of reusable parts? Furthermore, what about those miniscule bits of us whose value hinges on the encoded genetic information embedded within? Who are the rightful shareholders? Should it be us, individually and alone, or should we share ownership with others who are deemed genetically similar?

As this line of questioning illustrates, commodification is ultimately a transformative process. When Marx wrote of the strangeness of commodities, he was especially troubled by the inevitable mystification of a commodity's "origins," meaning specifically the exploitative labor processes that engendered its production. I, too, am deeply interested in the question of origins, although my intent here is to go well beyond an interrogation of labor practices. Of particular concern to me, on the one hand, are the medicalized processes that transform human body parts into objects of intense desire and, on the other, the subversive social responses that ultimately challenge the mystification of commodified organs' origins. More precisely, when body parts are given anonymously, culled from the dead, and then transplanted into ailing patients, involved lay parties may experience an intense longing to trace their exchange and, thus, track down one another. As I will illustrate below, the heavy medical reliance on the donor body as a source of transplantable parts may mean that involved lay parties, whose members began as strangers, may long to establish bonds of intimacy. Encounters between these parties circumvent professional and medicalized practices that commodify body parts and relegate patients to the generic category of "organ donor."

This second chapter, "Body Commodities," builds on the material discussed in the first, so I wish to revisit, very briefly, a few key points to underscore the linkages at work. The realm of organ transfer offers interesting examples of how we think about the human body and its reusable parts in this society. My comments here concern professional versus broader social responses to the culling of parts from dead bodies, or what I refer to as cadaveric donors. Organ transfer is particularly well suited to discussions of body commodification because standardized clinical practices, policy guidelines, and legislation concerning rights of body ownership and access were set in place forty to fifty years ago, specifically in response to the growing success of organ transplantation.

A factor of great significance is that cadaveric organs are almost always given anonymously in the United States. That is, details of donors' identities are not disclosed to patients when they receive transplanted parts. An important theme that will emerge later is that mandated anonymity encourages intriguing—yet unintended—consequences, engendering new forms of sociality between the surviving kin of organ donors and organ recipients. When organs are transferred from an anonymous, dead stranger, transplant recipients must ultimately imagine who their donors had been. All involved lay parties—including organ recipients and the surviving kin of deceased donors—are both moved and perplexed by the overwhelming evidence that death can beget life.

There is also an assortment of professionals whose work typically overlaps with the experiences of organ recipients or donor kin. At one end are transplant surgeons, nurses, and social workers; at the other are procurement specialists, who promote organ donation, acquire consent from the kin of dying patients, and might also sometimes assist surgeons in extracting usable parts from donors' bodies. Organ transfer's success is understood by this array of professionals as hinging on strict rules of decorum, where dominant ideological principles guide language, thought, and behavior.

As I noted earlier, in the United States transplantable human organs are understood as rare and precious goods, acquired through the generosity of donor kin to assist strangers whose very lives depend on their receipt. Perhaps the most important principle at work is that organ transfer relies exclusively on the "donation" and not the sale or forcible taking of human body parts. Donation is so intrinsic to organ transfer in this country that it is regularly referred to as "giving the gift of life," an act that is understood as helping others who might otherwise die from organ failure. All involved parties also emphasize that transplantable organs are in scarce supply, and this is asserted regularly and convincingly in the statistics quoted in transplant literature, by the media, and during transplant-related events. As noted in chapter 1, less than one third of those awaiting transplants actually receive matching organs each year, and, as UNOS and other organizations now estimate, approximately sixteen patients die each day waiting.[1] These grim figures lie at the heart, so to speak, of anxieties over organ scarcity in America. As also noted in the previous chapter, transplantable organs are at times even described as precious national resources.

When we speak of organ transfer as "donating life," or of organs as "precious national resources," we rapidly transform parts of ourselves into commodified objects of desire. But why—and how—do we allow such transformations to occur? One factor is our aversion to speaking openly about death. When we refer to the heart, lungs, liver, kidneys, and intestines as "gifts of life," we avoid dwelling on the horrors of massive head traumas and brain death; the deep sadness of sudden tragedy; the frustrations of hospital-based deaths; and the surgical realities involved in removing organs from the torsos of donors. As such, the language of the gift economy mystifies key aspects of organ transfer.

Among the most difficult topics to broach with those intimately involved in this realm is whether in commodifying bodies, we commercialize them, too. Clinical medicine is rife with specialized language, ranging from technical jargon for tools and techniques to slang that debases human patients to euphemisms designed to soften the emotional blow of devastating news (Coombs et al. 1993). Organ transfer professionals likewise draw on distinctive terms and turns of phrase in describing aspects of their own work. In the interest of exploring the paired themes of commodification and commercialization, I will focus briefly on euphemistic ways of talking about how transplantable body parts are procured in America.

Powerful forms of rhetorical policing (or what Richardson has dubbed "semantic massage" [Richardson 1996]) simultaneously underscore the life-affirming aspects of organ transfer while silencing discussions of body commodification. Consider this: if organs are donated as "gifts of life," how can we speak of a commerce of the body? Are not such objects in fact offered willingly by kin with the understanding that this might save several lives at once? As we shall see, the language of the gift economy is bolstered by national legislation, because the marketability of the body is highly dependent on how parts are categorized medically, socially, and legally in this country. Let us then consider aspects of the human body trade as a means to understand specifically what happens with transplantable human organs.

LEGAL ASPECTS OF THE HUMAN BODY TRADE IN AMERICA

The commerce in parts reaped from human bodies has expanded in extraordinary ways in recent years. Whereas half a century ago blood products figured prominently, our reproductive capacities are now marketable, too, such that financial compensation may be offered for sperm or eggs or surrogacy services. The reasons driving the legal commodification of these and other body parts over others are somewhat obscure and contradictory. One common argument is that it is permissible to market replenishable parts of ourselves because we incur no harm in doing so. Further, these replenishable products are in endless supply. As a result, laws permit us to sell our hair, blood, plasma, and sperm to various agencies around the country.

Ova, on the other hand, are always described as "donated," rather than bought, sold, or marketed. They stand in contrast to other body products in part because of their finite quantity. A woman's egg supply is understood as limited to the number with which she was born and, further, is regularly described as deteriorating in quality as she ages. Also, unlike sperm or blood harvesting, the procedures necessary for egg extraction are potentially dangerous. Egg "donors" are nevertheless understood as being "compensated," but never paid, for their time, commitment, and effort. To pay a woman for such "labor" would debase the assumed Samaritan act of assisting an infertile woman in need (Ragoné 1994, 1999). Such rationalizing nevertheless raises questions about the customary practice of compensating men financially for their reproductive capacities, whereas women are exempt (beneath the guise of protectionist rhetoric, at least) from the open marketing of their bodies (and their parts).[2]

Similar restrictions apply to whole organs. Although organ donors are sources of great medical worth, the value of their donations is clouded in a highly specialized language that privileges the gift economy and obscures references to commodification. This is especially evident where reusable human organs of cadaveric origin are concerned. Nevertheless, those who work for OPOs regularly quantify the worth of individual bodies, speaking of the "seven" or even "nine organ donor" as the quintessential success case. The donor body's value is similarly quantified by UNOS and other organizations, who regularly report that a single dead body may generate fifty or more reusable parts.[3] This figure has climbed steadily over the last few decades as clinical medicine finds even more ways to reuse the dead (Feher 1989; Flye 1995; Hogshire 1992; Kimbrell 1993; Machado 1998; Murray 1996).

Transplantable organs define a special focus in legislative initiatives that carefully restrict the human body trade in America. Most importantly, it is illegal to buy and sell transplantable organs. In response to an enterprising broker's attempts to offer kidneys on the open market, the Uniform Anatomical Gift Act of 1968 criminalized the selling of human organs for profit. (For a systematic review of amendments and drafted revisions, consult CUSL 1980; 1988.) As a result, all firms involved in the oversight, procurement, and redistribution of human organs are nonprofit groups, and these include the nation's OPOs as well as UNOS, which directs, on a daily basis, all OPOs' activities. The charges associated with organ transfer are understood as covering donor management, organ removal and processing, transportation, and administrative costs. They are not for individual organs themselves.

There is no denying, however, that organ transplantation is among the more lucrative medical professions in the United States and the accomplishments of a transplant unit ensure much local prestige for any hospital. Transplant surgery is also exorbitantly expensive, with an individual patient's costs ranging from tens to hundreds of thousands of dollars, depending on the organs involved and potential clinical complications. Yet transplant patients often link the range of costs incurred directly to certain kinds of organs over others. Professionals may even willingly do the same: in one instance from my research, patients based in a particular transplant ward were handed price sheets that detailed the differential costs for kidneys versus livers or hearts as part of the informed consent process (Sharp 1994). At least at the transplant end of organ transfer, it is well known that "donated" organs bear heavy price tags.

Other contradictions emerge when we consider human tissues—a category that includes corneas, skin, bone, and ligaments. In such contexts, the question of whether the body is commercialized is even more complex. Tissues are retrieved in tandem with organs and they, too, are donated by kin, who give their consent for both categories of body parts at the same time and often on the same release forms. Legal loopholes play a large part in exempting tissues from the restrictions imposed on whole organs, and thus they are just as likely to be handled by for-profit as by nonprofit groups. In fact, we readily acknowledge that tissues are commercial artifacts because we refer to involved agencies as bone, skin, and eye "banks." Economic survival as a nonprofit tissue firm is proving increasingly difficult: as the staff of one agency explained to me during a recent on-site visit, they are now considering shifting to for-profit status so that they can survive financially in a market now dominated by commercial firms.

Regardless of which body parts are acquired from the dead, charges that easily range between five and six figures are passed on to the consumer (that is, to the transplant recipient) or to their insurance provider. A dominant premise asserted in all quarters, however, is that issuing direct payments to surviving kin for the deceased's body parts threatens an already precarious realm of medicine that relies so heavily on selfless acts of human kindness. It also debases the sanctity of the body. The thought that financial compensation for body parts would quickly degenerate into sanctioned forms of blood payment, engendering exploitative practices that would prey primarily on the desperately poor, is also deeply abhorrent.

Nevertheless, anxieties over organ scarcity regularly spark debates on whether surviving donor kin should be compensated somehow. When I first entered this field of research in 1991, innovative reward proposals were generally dismissed immediately because they defied dominant assertions that linked body integrity to a respect for the dead. Professionals collectively expressed open disgust for blatant forms of body commodification, a stance guided in large part by the medical imperative to do no harm. Today, all involved parties embrace the need for ever more diligent public education campaigns, yet within the last six years financial incentives have surfaced as the most popular proposed solutions. The momentum of this approach is evident in recent policy statements issued by the American Medical Association, which, in partnership with UNOS, advocates investigating the viability of financial rewards (AMA 2003; Delmonico et al. 2002). Proposals now discussed frequently and openly at transplant events include offering donor kin a maximum $10,000 tax credit, a funeral expense supplement, a charitable donation credit, or a direct payment.

A dominant assumption driving these proposals is that Americans will commit more willingly and eagerly to organ donation when offered financial rewards. Proponents insist that such an approach is ethical because it will save more lives by alleviating the organ scarcity crisis. All realize, however, that such proposals require delicate public handling: the fear is that such policies will be interpreted as attempts to purchase human organs directly from the living or from surviving kin of the deceased. Such efforts risk being interpreted as nothing less than offers of blood money, especially when economically disenfranchised parties are involved.

Some of these recent policy initiatives nevertheless have shaped new state laws, and Pennsylvania (which houses several of the nation's most prestigious transplant centers) provides a case in point. State Act 102, which became effective in March 1995, was instigated by the late Governor Robert Casey, who was himself a liver and heart recipient. Act 102 offers a comprehensive approach to public outreach and donor motivation, strategies that other states and federal agencies have since duplicated. Yet another earlier and more controversial component involved offering partial financial assistance to donor kin for funeral costs (Nathan 2000). Although this section of the act failed to win full support in the legislature, at one point it generated a serious proposal for a three-year pilot project designed to test the effectiveness of $3,000 in funeral assistance to donor kin. If implemented, it would have been the first instance in this country of providing direct and legalized financial compensation to surviving family members. Enthusiastic consideration of this experimental strategy in some quarters marks an important watershed in proposals generated in response to anxieties over organ scarcity. Recently, Wisconsin passed legislation that reimburses *living* donors for travel expenses and lost wages, an example other states are certain to follow.

When I ask donor kin if financial incentives would have eased their decisions, they readily fall back on the language of the gift economy as a means to frame their responses. They assert regularly in interviews that financial rewards would debase organ donation and its associations with Samaritan acts of kindness. Such sentiments are especially pronounced among parents who have consented to organ donation when faced with their children's tragic deaths. In short, donor kin are far less concerned with *reimbursement* than they are with forms of *remembrance*.

Donor kin are also (sometimes painfully) aware of the ways organ transfer commodifies human bodies, and especially how it redefines known individuals as sources of reusable parts. This form of depersonalization is an unfortunate yet inevitable part of cadaveric donation, largely because organs are donated anonymously. Both procurement and transplant professionals regularly stress that anonymity is central to organ transfer's success. It allows donor kin to mourn their losses privately, without complications that might arise from exuberant or nosey recipients. Furthermore, too much knowledge of the donor might initiate pathological behavior in a recipient who is already deeply troubled by what some label "survivor guilt," or else by the sense that part of a dead person now resides within them. Transplanted organs should be viewed simply as replacement parts; also, the donor is *dead* and thus is no longer there. As such, the transplanted organ can harbor no trace or memory of its origins.

As a result of these professional understandings, it has long been considered taboo within the realm of organ transfer for donor kin and the recipients of their loved one's organs to be in direct contact with one another. In order to comprehend this longstanding policy we must look back—albeit briefly—at earlier cases of surgical experimentation that eventually gave rise to this successful realm of medical practice.

THE PRIVACY MANDATE

In the 1950s, with the first transplants (of kidneys between identical twins), organ transfer was an open affair. Within a few years, Christiaan Barnard's first attempts to transplant human hearts were widely publicized; one only needed to read *LIFE* magazine or watch the evening news to learn the identities of recipients and donors. Surgeons felt that all involved parties—from recipients to living donors to the kin of cadaveric donors—were healed emotionally if they were cognizant of one another's identities. A cascade of failures, all too frequently involving recipients' deaths, soon led to a much stricter, protectionist stance that rapidly evolved into policies that insisted on anonymity, at least where cadaveric donors were concerned (Starzl 1992).

At the beginning of my research during the early 1990s, I rarely encountered recipients who had met their donors' kin. Such encounters, when they did occur, were quietly facilitated by sympathetic transplant and procurement professionals. Involved parties understood they had broken official rules of decorum and were careful about when and with whom they spoke of these newly formed relationships. But much has changed, especially since the late 1990s, so that many OPOs now employ part- and full-time employees whose duties include overseeing correspondence between donor kin and their respective recipients. They also increasingly help orchestrate face-to-face encounters.

Reasons that have led to greater professional willingness to facilitate these meetings are complex (see Sharp in press); I will highlight only a few here. Transplant professionals have long been aware of the difficulties many recipients experience in accepting their "gifts of life" (Fox and Swazey 1992). An especially troubling yet widespread sentiment expressed by recipients is that someone had to die so they could live. Whereas transplant professionals stress that hearts, lungs, kidneys, and livers are simply replaceable parts, recipients are nevertheless frequently plagued by what Renée Fox and Judith Swazey refer to as "the tyranny of the gift." How are they ever to reciprocate gifts as precious and unusual as these?

At the donation end of organ transfer, we find that many professionals regard surviving donor kin as unpredictable and emotionally volatile. Professionals fear that if kin were allowed to meet the recipients who specifically harbor *their* loved one's organs, they might demand money or something else in return, cling furiously to their bodies, or weep profusely and uncontrollably. (I must stress that I have never witnessed any such responses.) As I described in chapter 1, these and other related professional anxieties shape official or bureaucratic memorial forms that commemorate the act of organ donation as a great social good. Participating donor kin are deeply troubled by such events because policies of imposed anonymity erase individual donors' identities. In response, many donor kin currently engage in alternative and thus subversive acts of memorial work, among the most prominent of which is publicly proclaiming the personal names of dead organ donors. This subversion has also played a large part in the gradual increase in face-to-face meetings between donor kin and transplant recipients linked through the same donor.[4] More enlightened professionals now lend their support as the gatekeepers of personal records, helping to assure a steady flow of written correspondence and later, perhaps, personal encounters.

PROTECTING THE HUMANITY OF HUMAN BODIES

As I elaborated in the first chapter of this work, organ transfer's success hinges in part on a broader tendency within biomedicine to depersonalize patients. As they shift to the status of organ donor, they are transformed from known individuals with personal histories to bodies that harbor precious commodities. The labels employed by professionals are especially revealing, because they reflect how involved parties learn to think about so unusual a category of patient. Brain death criteria, which figure centrally as determinants for organ transfer in this country, are central to the process of depersonalization. When patients are declared brain dead, they are rapidly (and even immediately) relabeled as "neomorts," "living cadavers," and the "half dead" by surgical staff who desire their parts; and as "donors" by procurement professionals who oversee their care exclusively once brain-dead status is confirmed. Such forms of objectification help rob patients of their humanity, because these specialized terms deny the presence of an individual residing within a given body (Hogle 1995; Lock 1996, 1997; Sharp 2001).

Nevertheless, we must acknowledge how intrinsic this process is—especially on an emotional level—to performing duties successfully in the realm of organ transfer. For instance, it is difficult to imagine surgeons or procurement professionals sleeping soundly if they thought they were taking organs from living human beings, were haunted by images of destroying the body's integrity, or dwelled too long on the idea that organs have been removed from the bodies of those whom kin know and cherish. From a professional standpoint, the social acceptance of organ transfer mandates that an individual, named patient be relegated to the new, anonymous, and ultimately generic status of organ donor.

This does not mean, however, that surviving donor kin and transplant recipients unquestionably embrace such (re)constructions of the donor self. As I will illustrate below, lay parties regularly reassign original identities back to deceased donors. When donor kin meet the recipients of the organs donated, they share their stories, ultimately engaging in a process of reanimating memories of the dead. Personal encounters engender strong, intimate bonds, or sociality, between these parties. Most compelling is how the presence of the donor's transferred parts enables assertions of familial ties between strangers. As an anthropologist, I am intrigued—and even drawn—to what we refer to technically as forms of "fictive kinship," or the practice of embracing nonrelatives as kin. In the realm of organ transfer this is possible when donor kin and recipients share the understanding that some essence of an individual donor can live on in the body of another. These are highly radical and thus subversive acts because they defy professional understandings that have inevitably commodified the donor body.

BETWEEN THE LIVING AND THE DEAD

"Commodities, like persons, have social lives," argues Arjun Appadurai (1986a:3). Marx, too, understood that coveted objects have social histories and asked us to consider what we might learn "If commodities could speak" (Marx 1967:176–77). What, then, does it mean to imagine commodified organs as possessing social lives? Should we really consider what we might learn if such coveted objects could speak? This strange line of questioning is certainly out of place when set within the strict confines of rational economic theory. But if transplanted organs are precious objects circulating within a national gift economy, do they not have social lives? Through the language of the gift economy they are, after all, no longer simply bits of flesh that work within their natal bodies; rather, through complicated technical endeavors, they now do their work elsewhere, inside other people. And through such complex rearrangements, transplanted organs take on new social meanings, especially among involved lay parties. It is, in fact, quite difficult for organ recipients to think of their newly acquired parts merely as sophisticated pumps or filters, or for surviving donor kin to imagine the dead as resting peacefully in the grave, when essential body parts have been redistributed to strangers living in diverse quarters of the country.

Regardless of professional assertions to the contrary, donor kin and organ recipients understand, in their own discrete ways, that transplanted organs are indeed like living things. After all, they bear the power to reanimate life in the bodies of those who were once sickly and dying. Whereas organ recipients speak of their experiences as a form of "rebirth," surviving donor kin embrace the idea that the lost loved one can "live on" in others. These paired sentiments are expressed frequently and openly during the Transplant Olympics, a biannual event hosted by the National Kidney Foundation that spans several days. Public speakers regularly refer to the bodies of the athletes, all of whom are organ recipients, as housing parts from donors who are imagined as participating in the range of competitions, too. Donors may even be described as fulfilling their dreams of being Olympians through the bodies that now house their transferred parts. (For an example of a donor's mother describing her deceased son in such terms, see Sharp in press, chapter 3). To the uninitiated, such a statement may seem strange. Yet these sorts of sentiments exemplify a highly naturalized way of speaking about the dead among donor kin and organ recipients.

Much can be gleaned from tracking the "social biographies" of commodified objects, argues Igor Kopytoff, for whom transplanted organs present an especially intriguing example. For the remainder of this chapter, I wish to examine the implications of this idea. As I will show, the medical act of transferring organs from the dead to the living initiates unorthodox ways of thinking about the "social lives of things" (Appadurai 1986b). Especially pronounced is the sense that organ transfer reanimates the dead: first, when the surgeon "sets" or restarts the heartbeat, or awaits the "pinking" or "pumping up" of the liver or kidney; second, through the inevitable biographical process initiated by intimate encounters between recipients and their respective donors' kin.

Within my first chapter I provided a sampling of visual images of memorial forms to show how donors are mourned in the transplant arena. Here I rely on the genre of storytelling to reveal how donor kin and recipients work together to initiate postmortem memories. By drawing from my field notes, I will illustrate how we might indeed track the social biographies of organs as precious, reanimated objects, imbued with the lives of their original owners.[5]

The time is a spring weekend in 1998; the location a landmark church in a booming metropolis on the eastern seaboard. The pews are peopled not with local parishioners but with scores of organ donor kin, who sit to one side of the main aisle; opposite them, on the other side, are half as many transplant recipients. All have gathered here on this blustery day to honor the donors whose bodies gave new life to transplant patients based locally or elsewhere in the nation. A range of speakers has been carefully selected through the combined efforts of a local procurement team, several hospital units, and a transplant patients' support group. Although many have come, essentially, to grieve lost lives, the tenor of the event is upbeat and referred to by at least two of the key organizers not as a memorial ceremony but as a "celebration of love and life." The altar is draped in purple in anticipation of Easter, a fact that inevitably leads one to draw parallels between Christ's resurrection and the idea that organ donation engenders rebirth in transplant recipients. A young man in his late twenties has just stepped down from the podium after speaking in moving terms of his brother, a firefighter who died in the line of duty, and from whose body two kidneys have granted new possibilities to a mother of two in her thirties and a retired grandfather in his sixties.

Norman Singer, a speaker who is well known in transplant circles here, is led to the podium by a young woman. Norman is a Euro-American man in his mid-forties with salt-and-pepper hair. He carries a white cane in one hand, and it is clear that he is legally blind. He begins in a quiet, even muffled tone, slowly growing more animated as his story unfolds. He speaks of his diabetes and subsequent kidney failure; his impending and progressive blindness; his struggles with dialysis three times a week; his desire yet failure to find full-time work because of his disabilities; and the associated financial and emotional hardships of all of these phases of his life.

Norman then turns to the first major transformation in his life: the night he received "the call" from his surgeon that a kidney match had been found, and that he should come to the hospital immediately so that he could be prepped for surgery. As Norman then explains in even more animated fashion, he always wondered about his donor, but for the first eight years he knew simply that this had been a young man in his twenties who, he later learned, was shot in the head by a stray bullet while crossing a major street during a police car chase in a neighboring city.

Norman pauses for a moment, reaches into his pocket, pulls out a fresh white handkerchief, and wipes tears from his face. He leans sideways and then returns to the podium; then he tells us, "I'd like to introduce my donor mom to you." (Note that he does not say "my donor's mom.") Mrs. Torres, a Latina in her sixties, rises from her seat, along with her adult daughter Rose, and together they join Norman at the podium. Mrs. Torres takes Norman's hand in hers and stands there quietly. As Norman explains, meeting this family is truly the most remarkable experience he has ever had in his life. Because of their willingness to help others when they themselves were faced with a horrible tragedy, Mrs. Torres and Rose have granted him, as well as three other recipients, new life. Most astonishing of all to him, however, is the way they have embraced him as a member of their family, such that he now feels he has not one mother but two, and a newfound sister as well. There is not a dry eye in the house, and before these three can step down, the attending audience applauds and then stands in a fashion most uncharacteristic for a church service or ceremony staged in honor of the dead.

In 1993, Harold Grimes, a Euro-American man in his mid-fifties, was diagnosed with an inoperable brain tumor. His physicians estimated that he might have only one more year of life because its location and rate of growth would inevitably cut off the oxygen to his brain and lead to brain death. As his wife, Judy, explained to me, Harold learned of organ transplantation when his brother underwent open-heart surgery. While visiting Harold's brother in the hospital, they met another couple whose son was there for a heart transplant; after learning what this entailed, Harold decided to spend the remainder of his own life preparing to be an organ donor. As his neurologist explained, because his tumor was not malignant, and because brain death was probably inevitable, it was certainly a conceivable plan. So from that point on, Harold exercised regularly, followed a nutritious diet, and made every other attempt to stay healthy so that his organs could be of use when he died. Harold and Judy went so far as to call their local OPO, which sent a staff member to their home to explain brain death and organ donation to them. As Judy recounted to me, the OPO employee even performed some preliminary blood work on Harold, and proclaimed, "You're a perfect candidate," and "I've never had the experience of talking to the donor before!" Harold made it clear to all intimate members of his family that his dying wish was for them to honor his plans to be an organ donor. (For a similar account involving a minor's planned death, see Greenberg 2003.)

When Harold was hospitalized for the last time, he refused all medications, even morphine (although he had been told it would not damage his organs). As Judy explained, "He could have died at home, but he chose to go to the hospital so he could be monitored and prepped for organ donation. We spent seventy days in the hospital waiting and watching him die. Each day they'd say, 'Well, he's close to brain death.' . . . He eventually developed pneumonia, so they couldn't use his heart and lungs. But his kidneys and liver were in good shape." Following several major seizures, Harold was pronounced brain dead on Valentine's Day.

Within a year of Harold's death, Judy and her daughter, Lizzie, attended a donor family picnic. It was Lizzie's birthday, but Judy had forgotten to get her a gift. Lizzie's only wish, however, was to learn what she could about her father's organ recipients, and, as it happened, an OPO staff member presented her with a letter from Scott Franklin, the fifty-five-year-old Euro-American recipient of Harold's liver.

They corresponded with Scott and his wife for over a year, at which time Scott invited them to travel 700 miles to attend a family reunion he was hosting in his back yard. Judy wrote several times in order to confirm that they were indeed welcome, and then she bought airplane tickets so that she, along with Lizzie and Lizzie's two young children, could attend. The event itself was held four years to the day after she buried Harold in his hometown.

As Scott explained to me, "We were giving out donor cards left and right! . . . It was such a wonderful weekend—I had 100 people to my house for the picnic, and even the local press showed up! We played [all sorts of games], you name it! It was like we'd known each other for years. Like a family I'd always had."

Although Scott and Judy live four states apart, they are now dear friends. Judy jokes around easily with Scott because he reminds her so much of her husband Harold. She sees many parallels, too, in Scott's own fight to get on a transplant list, and his long-term hospitalization following a complicated surgery. (He remained hospitalized for nearly a year.) As Judy explained, "When I first met Scott he was such a crack-up—he'd tease me in the same way that Harold would, in a way that no one else ever did. I had to tell him, 'Scott, this is really weird for me!' But when he says he'll stop, I tell him, 'No, there's no need to! It's just going to take some getting used to.'" As Lizzie explained, "My kids took to Scott like glue, and usually they're hesitant to approach strangers. Scott's like a big cuddly grandpa to them, just the way my dad used to be." When I ended my joint interview with them four hours later, Scott, Judy, Lizzie, and Lizzie's two children posed together for what they referred to as a "family portrait." The ease with which they sat together and leaned into one another's shoulders belies the fact that, contrary to appearances, Scott is neither Judy's husband nor Lizzie's father but, rather, the recipient of a deceased loved one's donated body part.

HIGH-STAKES CLAIMS ON DONOR BODIES

All of the people described above struggle on some level with questions about how to cope with the paired issues of loss and donor ownership. What rights do surviving kin have to trace the remains of their lost loved one? Can they assert claims of access, for instance? How socially invasive is this desire? And how much of the donor's tale should be revealed to recipients during face-to-face encounters? A shift to the recipient's perspective encourages a different yet related set of questions. Where lies the boundary between self and other? Should recipients squelch any sense that some essence of the donor persists within them? If they do embrace the idea that the donor is more than a mere body part, what and where precisely is the donor now?

In each of the stories I offer here, the donor assumes the form of a transmigrated, though partial, soul, inevitably generating compelling dilemmas for a range of involved parties. Are donors mere phantoms for donor kin, their memories fed only by past experiences? Can—or should—recipients integrate their imagined donors as parts of their now repaired selves? Face-to-face encounters simultaneously feed these imagined possibilities and confound them, especially when the involved parties become familiar with one another's tales. The creative playfulness of what we call fictive kinship enables these once disparate, now intimate parties to reconstruct together the trajectory of the lifetime (and postmortem) experiences of the donor.

I must pause to stress that I do not advocate that encounters should occur. Rather, I am interested in what happens when they do. It is worth noting that the data generated from my research contradict an assortment of professional fears. Many who work in the transplant arena regard bonds of fictive kinship as pathological because they can so easily imagine the emotional dangers of social intimacy. One cannot help but respect those gatekeepers who strive to protect the psychic health of recipients or donor kin; at risk is the further shattering of each person's world. However, so paternalistic an attitude may infuriate recipients and donor kin, for these individuals have already endured exceptional levels of professional intrusion in their lives. Yet as the small sample of successful encounters reported here reveals, potentially each party may be partly healed through repeated contact.

At this point, then, I wish to explore the theoretical implications of the encounter as a social process. From an anthropological perspective, how might we understand the cultural logic that drives donor kin and recipients to embrace one another as family? Of particular concern to me is how understandings of "blood ties" activate bonds of sociality. Organ transfer is particularly intriguing because the quite literal sharing of fleshy body fragments so quickly enables recipients to become the intimates of surviving donor kin.

GENETIC SENTIMENTS

The concept of kinship has long defined a central interest to anthropologists; one may go so far as to argue that an approach long referred to as the "genealogical method" (Malinowski 1961 [1922]) provides the bedrock for the bulk of field-based studies conducted throughout much of the twentieth century. It is, after all, through the study of kinship that anthropologists often decipher localized understandings of inclusiveness, difference, sentimental attachments, and animosity. American kinship is extraordinarily complex because of the diversity of forms that characterize so multicultural a society. Nevertheless, various authors have sought to identify dominant patterns and associated principles not only evident in everyday practices but also recognized in legal codes that ultimately shape conduct and the rights of individuals to associate with one another.

The writings of David Schneider stand out as classics in their attempts to identify such principles; among his more enduring arguments has been what he referred to as the assumed "biogenetic" quality of American ideas of relatedness. As Schneider explained, we understand our origins in such terms, where mother and father (or genetrix and genitor) share equally—or give equal parts—in creating a child. Unlike in other societies, where, for instance, children may be understood as being members of their mother's (matrilineal) or father's (patrilineal) kin, within the American context, children belong equally to both sides (what anthropologists refer to as a bilateral system of kinship reckoning). More importantly, their origins are understood in biogenetic terms, and the self is thus defined equally specifically through blood ties to both the mother and the father. This biogenetic quality is understood as inherently "natural" and in turn shapes bonds of kinship. That is, we are assumed to express especially strong emotional attachments for those to whom we have biogenetic ties. This idea of origins is exhibited in a range of contexts in our daily life: bureaucratic forms, for example, regularly ask us to name our mother and father, who ultimately are understood as defining our origins and thus, our identity. Unlike with nonbiological bonds defined through marriage, we cannot "divorce" our mothers or fathers. Thus, as Schneider explains, blood is a "natural, genetic substance" of connection that begins with birth and persists throughout our lives (Schneider 1980 [1965]:24, 1–27; see also Schneider 1984:165–77). As I will argue below, the concept of shared blood offers an intriguing framework for deciphering assertions of kinship in the specialized medical arena of organ transfer.

This form of biological determinism pervades American society, and is regularly and unquestionably embraced as scientific truth. Yet Schneider also underscored the flexibility of genetic reasoning within the context of American kinship: as new models or theories of genetic relatedness emerge, principles that guide how kinship is reckoned are inevitably integrated as social facts. Consider, for instance, our current fascination with the new genetics: descriptions of "blood ties" are no longer about uterine origins but include current understandings of how individuals share and pass on genetic properties. In fact, once unknown ties can now be asserted through DNA testing, a practice that, on the one hand, may problematize historical assumptions (consider recent debates concerning the descendents of Thomas Jefferson and the enslaved concubine Sally Hemings), and, on the other, may assist kin in retrieving the remains of soldiers missing in action from the Korean and Vietnam wars. In other words, blood relatedness is currently conceived of as operating at the cellular or genetic level. Of particular significance to my arguments is Schneider's insistence on the *sentimental* quality of blood ties, such that bonds of kinship may be understood as forms of "instinctive affection" (Schneider 1984:167).

I am interested in how Schneider's concept of biogenetics proves relevant to exploring new forms of sociality in contexts framed by organ transfer. If we consider the fact that there are neither shared offspring nor conjugal unions bridging the transplant divide, how do donor kin and recipients assert ties of relatedness? In what ways does the transfer of body parts translate into ideas of sharing, or sociality among disparate parties? How do related sentiments confirm, challenge, or extend Schneider's ideas from twenty-five years ago?

If we compare organ transfer to other domains of science, we encounter radically new ways of thinking about the sentimental nature of biological ties.

In considering similar quandaries in other medicalized quarters of daily life, yet another anthropologist, Paul Rabinow (drawing on the work of Michel Foucault) has argued recently that the new genetic technologies foster new forms of "autoreproduction." Consider, for instance, how we understand the concept of family in reference to prenatal genetic testing, surrogacy, and cloning. Reminiscent of Schneider's concern for biogenetics, Rabinow (like Foucault) employs the concept of "biosociality" to flag the manner in which medical technologies may reshape social processes (see Foucault 1978; Marcus 1995; Rabinow 1992, 1996). Anthropology includes a burgeoning literature that illustrates the pervasiveness of this type of thinking in reference to such diverse experiences as gestational surrogacy, children with Down syndrome, and various reproductive technologies where both humans and animals are concerned (Franklin 1991, 2003; Ragoné 1996, 1999; Rapp 2000). Especially intriguing is how the sharing of blood (and genetic characteristics) offers a simultaneously compelling and flexible means for expressing similarities among strangers.

Whereas involved actors may celebrate these new forms of sociality, Rabinow warns that "the new genetics will carry with it its own distinctive *promises* and *dangers*" (Rabinow 1992:241–42) (italics mine). The realm of xenotransplantation (which involves attempts to merge human and animal species) certainly seems to herald such warnings (Sharp in press). (I will return to this topic in my third and final chapter.) Of particular concern to Rabinow is the power of medical practice and knowledge to reshape seemingly mundane aspects of our social worlds (Rabinow 1992:241–42). I am nevertheless intrigued by the potential of burgeoning forms of biosociality to transform social relations among strangers. As the stories offered above reveal, encounters between former strangers engender a new and deep sense of belonging to one another, such that recipients and donor kin are redefined as mother and son, brother and sister, widow and surrogate spouse, or grandfather and grandchildren. In such relationships there is renewed comfort, rendered possible only by the fact that a deceased loved one's body parts now enable other people to survive.

With Schneider once again in mind, we see from such stories that blood ties may be deeply sentimental. In the realm of organ transfer, the sharing of body fragments generates not only a strong sense of sameness but also strong emotions. Organ transfer is unique in its ability to transcend normative (or natural) unions. The notion of sameness is at once about shared human fragments, sentimental bonds of love, and denying the commodification of the body and its parts.

In conclusion, if we return to Marx's question—What if commodities could speak?—we find that transplanted human organs are perceived by some parties as living things. They grant new life to recipients, who consider their personal surgical transformations to be a radical form of rebirth. In turn, a donor might well be perceived by surviving kin as having been reborn in the bodies of organ recipients, and spoken of openly and plainly as living on in several strangers. These paired rebirths generate unorthodox yet highly creative ways to confront grief as well as subvert the medical commodification of the human body.

The process of reanimating the dead is no trivial thing, though, for the new forms of sociality that emerge through face-to-face encounters generate their own dilemmas. The sense of longing, expressed in various ways by donor kin and transplant recipients, offers a final case in point. Kin may voice the strongest desire to meet the heart recipient, which exposes how a hierarchy of value is assigned to the body (or, at the very least, to its transplantable parts). In turn, recipients overwhelmingly place the highest value on donor mothers, so much so that awards are given regularly around the country in their honor. (A recipient's desire to encounter his or her donor's mother, I posit, may spring from the fact that each has harbored some essence of the donor inside their own bodies, one through surgery, the other through pregnancy; see Sharp in press.)

These are some of the more startling consequences of cadaveric organ transfer. As involved parties struggle against rules that dictate donor anonymity, their reunions ensure that donors may "live on," so to speak, in the hearts and lives of others so intimately connected through the literal sharing of flesh and blood. But what of even rarer instances of body melding, where human patients must integrate nonhuman parts? What then might happen in the struggle to trace a commodity's origins? These are the sorts of questions I address in the third and final chapter, which concerns the scientific desire to create new hybrid forms by melding human bodies with parts of animal and mechanical origins. As we shall see, these very real concerns characterize the brave new world of contemporary transplant medicine in America.

THREE

HUMAN, MONKEY, MACHINE

The Brave New World of Human Hybridity

THE BABY AND THE BABOON

In October 1984, Loma Linda University Medical Center in California made a startling public announcement: surgeon Leonard Bailey had removed the flawed heart of a four-day-old baby girl and replaced it with one from a female baboon drawn from Bailey's research colony. The infant—dubbed Baby Fae—reportedly fared well at first, but she soon succumbed to an onslaught of clinical complications associated with acute graft rejection. This form of organ transfer, involving a xenograft, attracted immediate and widespread media and medical attention, and responses ranged from amazement to anger, horror, and disgust. Although Bailey's research is now heralded in some quarters as having advanced clinical knowledge on newborn transplant surgery, he was openly chastised and later shunned by many colleagues for his medical hubris. The fact that Loma Linda was a Seventh-Day Adventist hospital, and that Bailey himself discredited evolutionary theory in at least one public interview, led to even further rebukes from a scientific community that questioned his motives as a marginalized and underfunded researcher (see ABC 1996; Goule

1981; Stoller 1990). Opinions were far from unified, however: whereas some found his use of Baby Fae as guinea pig despicably narcissistic, driven by a personal desire to promote what some deemed an otherwise mediocre career, others were dumbfounded that he would attempt a xenotransplant when previous trials (by high-profile surgeons) had ended rapidly in patient death without generating findings that would justify new attempts anytime in the near future (HC 1985; Altman 1984a, 1984b, 1984c).

Still another response focused more specifically on the nature of the animal itself. Whereas chemically treated, inert pig heart valves had already proved to be effective replacements for failing human ones, it was widely accepted that the human immune system would not tolerate the full-scale replacement of an organ with one taken from an animal. Such opinions were not only derived from ongoing laboratory experiments but also based on earlier failures to graft baboon and chimpanzee hearts and livers to human bodies (see Starzl 1992). Thus, some critics recoiled specifically from Bailey's use of the baboon, insisting that it involved the unwarranted sacrifice of a precious animal. In other camps, an underlying yet less clearly articulated anxiety was that there was something improper—no, unnatural—about Bailey's act, where two evolutionarily close yet distinct primate species were merged within one body, and an innocent baby at that. The fact that the transfer involved an inferior simian "cousin" struck some as especially repulsive or, at the very least, a source of apprehension about the medical melding of disparate bodies. The paired terms "baboon" and "heart" (alongside "baby") bore especially heavy symbolic weight.

Still, this was hardly an isolated event in experimental transplant medicine. While the case of Baby Fae is emblematic of attempts at xenotransplantation, bioengineering defined another front of research activity. The heart had long been a focus for experimentation, with mechanical innovations extending back to the first half of the twentieth century. Among the earliest technological successes sprang from the combined efforts of aviator Charles Lindbergh and vascular surgeon Alexis Carrel, who together designed a full-body perfusion device in 1935. This invention was the precursor of the contemporary heart-lung machine, which artificially powers the body's blood circulation during bypass surgery.

Whereas this now routine device serves as a temporary bridge for the heart, other efforts have focused squarely on full replacement. By the 1980s (coinciding with Bailey's work), efforts to replace the human heart were conducted in earnest, with the Jarvik-7 heart a major experimental breakthrough, and one attempted first in animals and then in humans (see figures 3.1 and 3.2). This device, however, was far from portable: although a mechanical pump replaced the excised natal heart, patients remained tethered to an apparatus the size of a washing machine. Barney Clark, who was the Jarvik team's first patient, died on March 23, 1983, after surviving 112 days; even more remarkable was "Bionic Bill" Schroeder, who died in August 1986 after "620 grueling days" (Fox and Swazey 1992) of living with a device whose persistent noise he compared to a threshing machine (Siebert 2004:261).

After a decade-long hiatus following the testing of the Jarvik-7, there have been more recent attempts to try out new heart prototypes, now involving fully implantable devices known as TAHs or Total Artificial Hearts, of which there currently exists a range of experimental models. Perhaps the best known today is the AbioCor, a device celebrated as an "Invention of the Year" in 2001 by *TIME* magazine

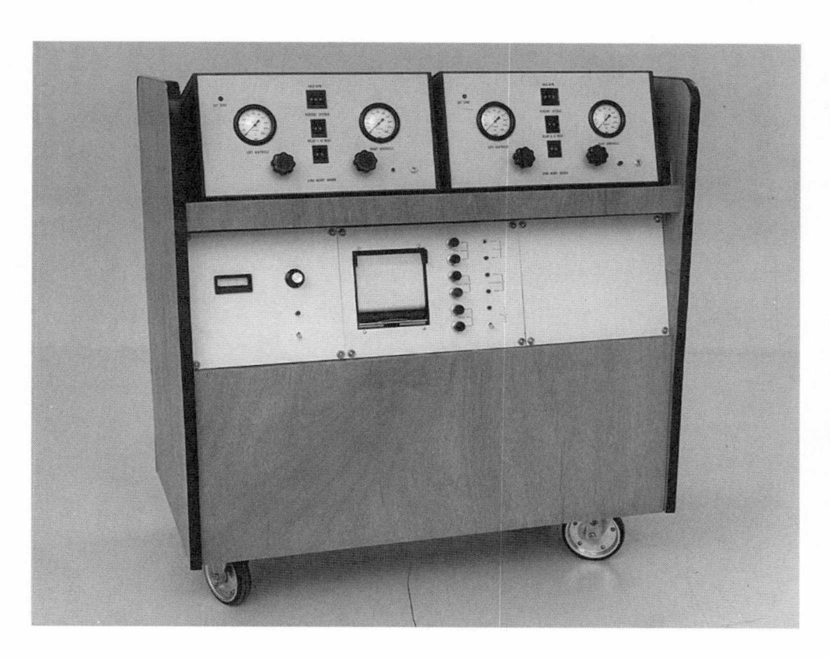

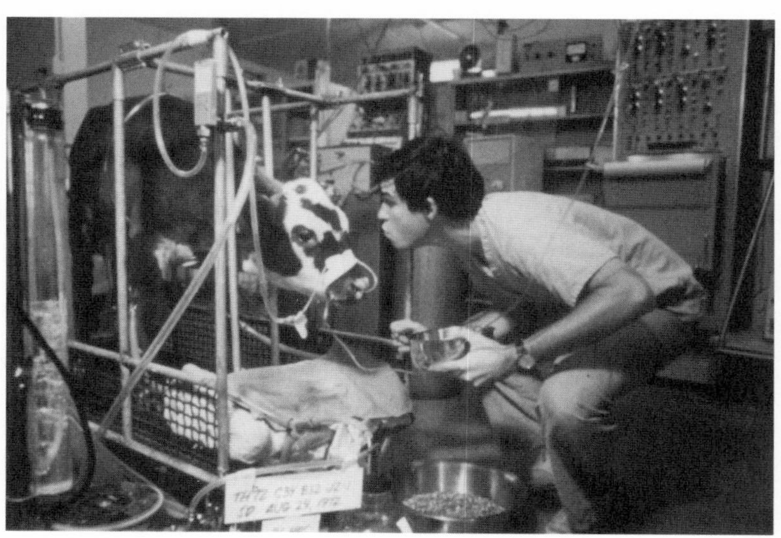

(Hamilton 2001) (see figure 3.3). This mechanical heart is the creation of a biotech firm based in Dedham, Massachusetts, which has been granted permission by the Food and Drug Administration to test the device in 15 patients. The first to receive it was Robert Tools, a 59-year-old African American man suffering from congestive heart failure as well as liver and kidney damage. Tools survived 119 days with this battery-powered, "softball-size" heart of plastic and titanium embedded in his chest. The record for the longest survival rate to date (16 months, 3 weeks, and 4 days) was set by Tom Christerson, a 71-year-old man who was the second of now 14 recipients of this heart (CHFpatients.com 2004; Fox and Swazey 2004; *The Courier-Journal* 2001).

These and related mechanical prototypes appear to offer the greatest promise as a bridge of sorts, as proved in September 2004, when surgeons at Stanford University temporarily connected five-month-old Miles Coulson to, in this case, an external Berlin Heart model while he waited for a human heart of proper size (which he subsequently acquired) (Richter 2004a,

Opposite page

FIGURE 3.1. (*top*) The Utah Heart Driver, the apparatus that powered the artificial Jarvik-7 heart that was first implanted in patient Barney Clark in December 1982. The gauges on each side read "Left Ventricle" and "Right Ventricle." *Special Collections, J. Willard Marriott Library, University of Utah.*

FIGURE 3.2. (*bottom*) The original caption to this photo reads: "Robert Jarvik in 1972 discussing with the calf the clinical results of his latest artificial heart with which this calf is provided." The animal was an early experimental recipient of the prototype for the Jarvik-7 artificial heart; Robert Jarvik remains a pioneer in artificial heart research. Photo from the personal scrapbook compiled by Dr. Willem Kolff, who designed the first artificial kidney and, later, the heart-lung machine. *Special Collections, J. Willard Marriott Library, University of Utah.*

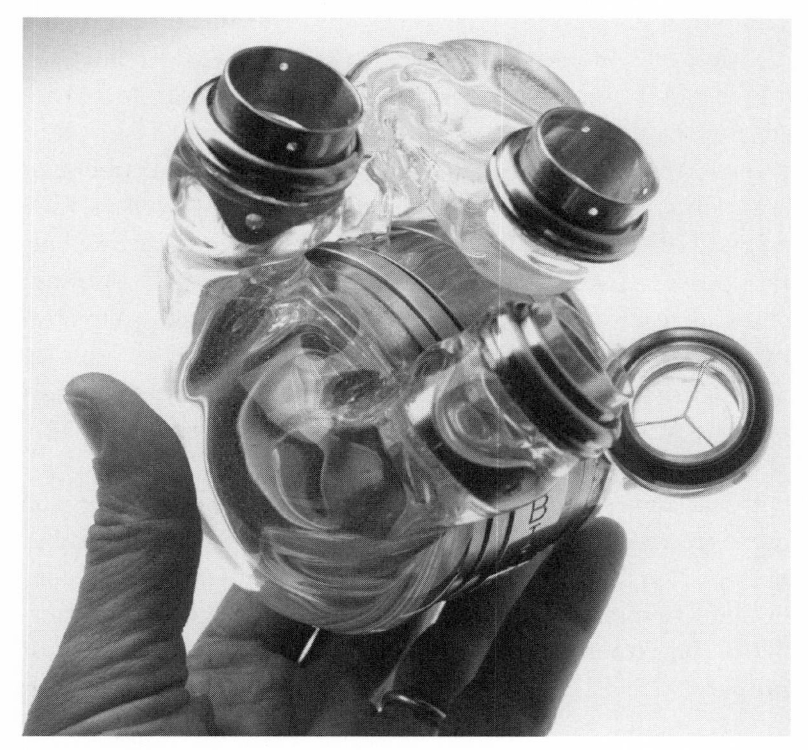

FIGURE 3.3. The Thoracic Unit of the AbioCor total artificial heart (TAH), a fully implanted device that replaces the recipient's original organ. It is powered by an internal rechargeable lithium battery and controlled by an implantable electronics pack. There is also an additional external battery pack. The design eliminates the need to tether patients to large consoles, as with the Jarvik-7. *Photo courtesy of ABIOMED, Inc.*

2004b) (see figure 3.4). For adult patients, however, the prospects have not been nearly as promising (in part because adults are more often asked to submit themselves to research knowing that death is inevitable). Although fully implantable devices offer the promise of greater mobility one day, human subjects who agree to TAH implantation are destined to remain hospi-

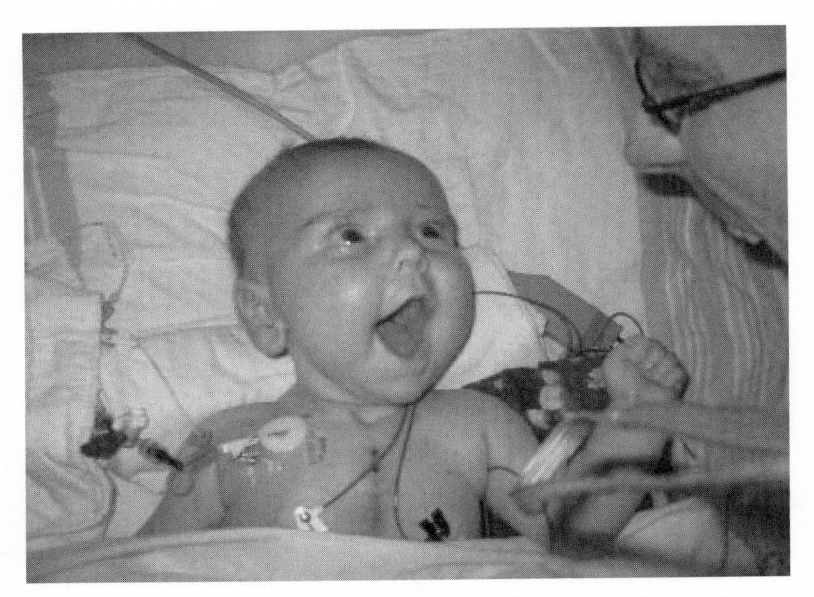

FIGURE 3.4. Miles Coulson, who was temporarily connected to a German (the "Berlin") heart pump, at five months in 2004. This fist-sized, external polyurethane device powered his ailing heart while he awaited an appropriate match for a human heart transplant. *Photo courtesy of Lucile Packard Children's Hospital, Stanford University, and the Miles Coulson family.*

talized, although brief forays outdoors are sometimes allowed. Experimental recipients endure constant medical scrutiny throughout the remainder of their lives; their survival depends on their being tethered to an array of machines that monitor their physical status and that of the heart device itself because it currently bears lethal consequences for recipients. Life-threatening complications include blood clots, uncontrollable bleeding, strokes, and multiple organ failure. Sadly, these and other lesser or unanticipated effects may prove unbearable for patients, as well as for kin who are helpless witnesses to the suffering and death. Such has been the case for at least one now deceased AbioCor recipient, whose family became embroiled in a complex legal suit against the patient's health care team (Fox and Swazey 2004). Far greater success has been achieved with implantable heart pumps that assist a failing (and intact) heart in performing efficiently. The most impressive example today is the life of Peter Houghton, who has survived over five years with a fully portable, battery-powered Jarvik-2000 Heart Pump. The device works so well for Houghton that he is now an avid hiker and globetrotter, after having faced death's door because of the debilitating effects of severe heart failure (Houghton 2001).[1]

BODY EXPERIMENTS

Today, these two experimental trajectories—one involving animals, the other biomechanical prototypes—are marked by a greater sense of urgency than twenty years ago, as transplant medicine faces the bleak reality that the demand for human parts far outpaces the nation's supply. This third and final chapter will focus on varied understandings of what are clearly competing experimental realms of transplant research in order to explore how a range of affected parties perceive them. I begin by considering the manner in which animal- and mechanical-based innovations herald scientific breakthroughs among researchers. This approach helps to expose the slippery aspect of how, in turn, the embodied potential of hybrid forms is imagined by the very people whose bodies may have to rely on such unusual forms of repair. My discussion here builds on the two previous chapters concerning the good death as framed by organ donation and the shared desire among organ recipients and donor kin to forge bonds of sociality. As I have underscored, organ transfer involves intimate forms of body melding. I now return to the theme of intimacy, this time in order to consider the social implications of melding human bodies with parts of various nonhuman origins.

As noted earlier, organ transfer in America is plagued with anxieties over the scarcity of transferable human parts. As transplant surgery expands as a profession and perfects even further its technological craft, procurement professionals experience ever-increasing pressure to supply the growing need for organs. Sadly, the demand far outnumbers the sum total reaped, and this will continue to be so unless we radically redefine the legal parameters of death, or expand the criteria under which organs can be taken from both the living and the dead.

Set against this medical reality, organs of animal and mechanical origins promise to alleviate the shortage and, potentially, even replace human parts altogether. At present, both of these potential, experimental paths face serious clinical obstacles: the *acute* or immediate rejection of xenographs is so pronounced that researchers are unable even to describe what *chronic* or long-term rejection or survival might look like; and no mechanical device is yet capable of replicating the precise workings of the human body over the long term. Organs derived from animals may harbor hidden pathogens that might jump the species barrier and be transmitted to a human patient, and then to his or her intimates and even to a wider community beyond. (For a case involving the transmission of rabies from a transplanted organ see AP 2004; CDC 2004; UNOS 2004). Also, ironically, such mechanical devices work too efficiently, causing a host of life-threatening problems in other regions of the body. Such medical hazards, however, do not prevent a range of interested parties from imagining the potential wonders (and dangers) (cf. Rabinow 1992) of alternative forms of human organ replacement.

My focus here, then, is ultimately the social imagination. I begin by exploring forms of scientific longing, looking specifically at how involved researchers speak of the potentials borne by their work and how they privilege certain forms of body melding over others. Just as organ transfer is plagued by anxieties over scarcity, researchers face the constant challenge of obtaining adequate funding. Because the financial stakes are so high (and the government commitment relatively low), leaders in the field regularly establish corporate partnerships as a means to fund their work through private investment. As an interlude of sorts, then, I will pause briefly to consider the current (though inevitably fickle) voice of venture capital as a means to underscore its deep fascination with technology and blatant (and even self-effacingly comical) disregard for human considerations. I then conclude by exploring responses offered by potential recipients—individuals who can imagine themselves as one day harboring parts of foreign origin within their own bodies. Each of these sets of responses is shaped not only by pragmatic considerations but also by the symbolic values assigned to various animals and machines. Although such sentiments may not be fully articulated by science and venture capital, the voices of organ recipients make it all too clear that the image of Frankenstein's monster may override a longing for pragmatic solutions in other quarters (Cohen 1996; Helman 1988, 1992). Nevertheless, within each of these domains—science, financial investment, and patients' lives—we encounter intriguing responses to the implications of sociality specifically in reference to truly radical forms of body melding, or what I will refer to as forms of human hybridity. I begin by exploring, from a historical perspective, the scientific desires within transplant medicine associated with melding humans with animals and machines.

THE SCIENTIFIC DESIRE FOR THE HYBRID HUMAN

Medical history has long included efforts to experiment with forms of human-animal hybridity. Blood transfusions from animals to humans were attempted in the early seventeenth century in England and France; during the nineteenth century, skin from a variety of animals was grafted onto human patients; and from the 1890s through the twentieth century, there were numerous attempts to transfer organs from pigs, monkeys, and apes to humans (Starzl 1992:113). Organ transfer today also owes its success more generally to experimentation on animal models: rodent, canine, and simian bodies have long served as prototypes before procedures are tested on humans. One need only consider, for instance, the professional trajectory of Thomas Starzl, the preeminent liver surgeon long based in Pittsburgh, to recognize the importance of xenotransplantation experiments.[2] As Starzl notes in his memoir *The Puzzle People*, his own attempts at xenotransplants date back to the 1960s, when his team in Denver transplanted baboon kidneys into six patients; the grafts lasted six to sixty days. In 1963, another surgeon in Minneapolis secretly performed a baboon kidney transplant into a sixty-five-year-old woman, and the case was made public only after yet another surgeon at Tulane attempted several chimpanzee kidney grafts (one of which functioned for nine months). In the same year, a chimpanzee's heart was implanted in the body of a sixty-eight-year-old man, who died within hours of the surgery. A similar fate befell a young woman who, in 1977, also received a simian heart (Starzl 1992:112–14).

The development of contemporary immunosuppressant drugs ultimately paved the way not only for improved patient survival rates but also for subsequent animal experiments. An important watershed in transplant medicine was the advent of cyclosporine in the 1980s. This potent immunosuppressant has enabled many transplant recipients to survive decades postsurgery. It also played an important role in the Baby Fae case, an event that is extraordinarily significant historically not simply because it involved an attempt at simian-human melding but also because such an attempt was assumed possible after this and related breakthroughs in drug therapy.

Although the experimental front remained relatively quiet for a decade, still other new drug therapies led transplant researchers to reattempt xenotransplantation, this time in Pittsburgh, a leading transplant center. In 1992 and 1993, surgical teams, working under Starzl, tested the effects of transplanting baboon livers into adult patients in conjunction with a drug regimen that included the new immunosuppressant FK 506 (Starzl et al. 1993). The patients involved were two men, thirty-five and sixty-three years old, both of whom were infected with active hepatitis B and for whom (at the time) there was no chance of a human match. Both received baboon livers.[3] These men survived seventy and seventy-one days, respectively, ultimately dying from complications associated with acute graft rejection and accompanying clinical treatments. This nevertheless set the stage for subsequent work by teams composed of immunologists and surgeons who began in earnest to explore cross-species grafting (where, currently, donor and host are both nonhuman).

As I detail elsewhere, surgeons and other transplant professionals overwhelmingly prefer xenotransplantation to mechanical replacements, whereas for potential recipients, the opposite is true (Sharp in press). I argue that professional preferences hinge on the fact that xenotransplants require both surgical and immunological expertise, a combined background typical of those who already work in transplantation, whereas mechanical devices fall within the domain of bioengineering. Although orthopedists are well versed in skeletal mechanics, cardiologists, hepatologists, and nephrologists rely on the expertise of other technicians to oversee and operate the range of machines that facilitate their work with patients. They are not in the business of designing mechanical prototypes. Furthermore, much of their surgical training (especially if they have spent time in a research lab) involves working with animal models, and thus the leap to xenotransplant experiments is a natural one, so to speak, for them. Xenotransplant researchers and surgeons regularly stress in interviews and less formal conversations that human and animal parts share "natural" or "fleshy" origins, and so they view these as inherently superior to what they think of as clumsy, rickety mechanical devices.

The professional favor shown for xenotransplantation was especially pronounced during a recent meeting I attended of the International Transplantation Society in Vienna in September 2004. While multiple sessions were devoted to xenotransplant research and plenary sessions hailed it as the future hope in the field, relatively little regard was paid to mechanical alternatives, even though a landmark case at Stanford University, involving the infant Miles Coulson, was unfolding that very week (again, see figure 3.4). One might conclude that such myopia sprang in large part from the profile of the society's current leadership, which includes a number of leading experts in xenotransplantation. In the course of my research in other venues, however, I find that surgeons largely prefer fleshy body parts to mechanical substitutes.

Prospective patients—those whose fates hang in the balance as they await news of an organ match—overwhelmingly express a preference for mechanical parts. Granted, such choices are made under hypothetical circumstances; nevertheless, when questioned in interviews, patients, regardless of gender or age, generally mistrust parts of animal origin. The most pronounced anxieties are driven by an aversion for animal-human forms of body melding. A frequently expressed concern is that the patient might take on the characteristics of the animal in question; when this is imagined, responses range from a desire to cut short the conversation to a tendency to degenerate into ribald joking behavior. The words of a Euro-American man in his forties, whose brother had had a kidney transplant, are fairly typical. When asked to consider various alternative organ sources, he responded that "it would be a little strange to have a baboon heart. Would I start baring my teeth and bottom?" (for this and other related examples see Sharp in press).

As described in the previous chapter, transplant recipients are all too familiar with the psychosocial complexities associated with body melding when replacement parts are derived from cadaveric donors. If human organs inspire creative forms of sociality between donor kin and recipients, what might transpire if a person's life were saved through the sacrifice of a baboon or a pig? Major players within the realm of experimental research are far more interested in the pragmatic side of their work than in the social or psychological dilemmas that could accompany successful breakthroughs. Potential patients are understood as passive recipients of groundbreaking prototypes and techniques that, although currently experimental, might well herald workable, life-saving approaches in the near future. Such sentiments are especially pronounced in various quarters that remain uninhabited by patients. In an attempt to make sense of the allure—and revulsion—associated with animal and mechanical alternatives, I consider first the scientific romance with fleshy hybrids, then the technological dreams of venture capital.

CELEBRATING THE MONSTROUS IN LABORATORY SETTINGS

Donna Haraway, the author of *Simians, Cyborgs and Women: The Reinvention of Nature,* has argued that practices inherent in late twentieth-century capitalist medicine are frequently characterized by "critical boundary breakdowns" that ultimately bear great potential for reshaping humanity. Emergent forms of hybridity include both "intensified machine-body relations" and, in other quarters, the thorough breaching of boundaries "between human and animal" (Haraway 1991:151, 171). Although Haraway is most concerned with non-transplant related forms of body melding (as exemplified by computer interface techniques, the new genetics, and primate behavioral studies), her argument that we might well celebrate the hybrid body proves especially helpful in understanding how research scientists conceive of xenotransplantation's potential.

Celebrations are indeed widespread in the realm of transplant medicine, although enthusiasm for hybridity is hardly uniform. There is, nevertheless, a pronounced preference for solutions that meld human bodies with nonhuman mammalian species. Among the more intriguing developments involves the growing ambivalence expressed about simian "donors" against a mounting desire to meld human bodies with organs of porcine origin (see, for instance, Fishman et al. 1998).

Transplant medicine is populated with a range of monsters, employing a language that at times condemns and at other times celebrates the biomedical potential of hybridity. As I mentioned in chapter 1, transplant patients who identify too strongly with their deceased donors may well be labeled with "Frankenstein Syndrome," because they suffer from the delusion that they have been pieced together with parts from disparate bodies (Beidel 1987; cf. Helman 1988, 1992; Youngner 1990:1015). Yet medicine has in fact naturalized other kinds of hybridity. For instance, a technical term employed with gusto is "chimerism." (Recall that originally, the Chimera was a monster in Greek mythology that preyed upon humans; it was part goat, part serpent, and part lion, and it was slain by Bellerophon, who rode the winged horse Pegasus.) In contemporary biomedical contexts, the term is used to describe the successful integration of immunologically distinct bodies or their sectioned parts (Antin et al. 2001; Jankowski and Ildstad 1997; Quaini et al. 2002; van den Bergh and Holley 2002). More specifically, within transplantation research, studies concerned with chimerism involve attempts to disable or deceive the human immune system so that parts (be they organs or miniscule bits of tissue) derived from other species are read as "self" rather than as an invading Other (BioTransplant and MGH 1999). Researchers hope to overcome immunological impediments so that they might one day even eliminate the need for immunosuppressant drugs (Chillag 1997:70–73). The most complex projects today involve forms of genetic tampering with donor species, whose DNA is encoded with human material. By crossing the species barrier in this way, researchers hope that transplantation might eventually rely on xenogenic (or trans-species) grafting of transplantable parts, thus eliminating forever the need for allogenic (that is, human-to-human) forms of organ transfer.

Truly remarkable chimeras are emerging in the realm of the new genetics. A recent wondrous, yet bizarre example is the spider goat. Although this creature looks perfectly ordinary, it has been genetically engineered to produce milk that bears tensile fibers previously generated only by orb-weaver arachnids. The spider goat attracted media attention in 2001 because of its sheer perversity (Helman 2001). The mere existence of this mismatched hybrid inevitably raises pointed questions about just how far science should go in altering the capabilities of a species. A far more difficult project in transplant medicine is how to integrate parts from another species in the mature (versus embryonic) body and, further, which animals offer the best human match.

OUR SIMIAN COUSINS

In the United States, evolutionary biology plays a significant role in determining which animal models are considered to be the most suitable human prototypes. For instance, apes and monkeys have long defined an important category of research animal in realms ranging from experimental psychology to primatology. From chimps and other simian species we seek to learn about primate cognition and social development and draw conclusions that may help us understand the workings of the human brain. A few research subjects have even traveled to space and back. Nonhuman primates are especially prized within American medicine because of their evolutionary proximity to us. More particularly, within the realm of transplant medicine, they are valued because of their anatomical similarity, and their organ size provides a convenient match for at least some human bodies. The extent to which we take this for granted is reflected quite clearly in the Discovery Channel series on primates, which produced a companion coffee-table book, *Cousins: Our Primate Relatives*, its cover featuring a chimpanzee scratching its head (Dunbar and Barrett 2000) (see figure 3.5).

Today, however, the scientific use of simians produces significant problems within the research realm. For one, they are extraordinarily expensive to acquire, breed, and maintain. This

FIGURE 3.5. Cover image from *Cousins: Our Primate Relatives*, by R. Dunbar and L. Barrett. *Reprinted with permission from DK Publishing.*

is in large part because strict guidelines protect their welfare as laboratory subjects. Furthermore, once they are no longer needed in the lab (and if they are not sacrificed at the end of an experiment), legal guidelines require that they join retirement colonies where they can be cared for until the end of their days, and monkeys and apes can live for decades. For merely pragmatic (and financial) reasons, other animals are regularly used instead, including rodents, dogs, and pigs. This is especially pronounced in transplant research, where among the most striking developments in recent years is what I will refer to as the scientific "romance of the pig."

FOR THE LOVE OF THE PIG

The desire to meld human and porcine bodies is rooted in older, established medical practices. As noted earlier, a now highly successful use of porcine parts involves the regular replacement of damaged human heart valves with those derived from pigs (Cooper et al. 2002; Maeder and Ross 2002). In reference to full organ transfer, much attention is focused squarely on interbred miniature swine. These transgenic creatures harbor human material that was spliced to pig cells during very early stages of reproduction. Transgenic pigs have been produced within laboratory settings for over twenty-five years (Cooper et al. 2002), a fact that underscores that such practices have indeed become routinized within certain experimental realms of science.

I think it is safe to argue that the current scientific value placed on porcine bodies is altogether peculiar from a lay perspective. As numerous anthropologists explain, pigs are regularly devalued in a variety of realms within our society as a source of filth, as is evident, for example, in the dietary prohibitions observed within a range of faiths (including Islam,

Judaism, and some Christian groups). A widespread assumption among many Americans, too, is that that pigs are messy creatures that prefer unclean environments (Douglas 1966; Harris 1985; Leach 1964). Yet pigs are also highly regarded by others for their intelligence or as sources of food.

At first glance, though, these culturally based understandings seem to offer little insight into why science is so set on melding humans with pigs. Partial answers lie in how science categorizes experimental animal species. The potential of the pig rests in large part with the immunological and physiological similarities it shares with humans. Its value is also rooted in the fact that it is does not rank as high in evolutionary terms as monkeys and apes. The pig emerges as a boon for transplant research because it is a domesticated farm animal, not a close cousin.

Other strictly pragmatic differences have enabled the pig to surpass the chimp or baboon as the experimental creature of choice. Pigs mature quickly and reproduce with ease, and sows deliver litters, not single offspring. Pigs are much cheaper to acquire and raise, and fewer regulations dictate how they must be cared for or "sacrificed" once experimentation is concluded. The price tag associated with acquiring a baboon, for instance, may be so high that the same animal may serve as a work object for an array of graduate student projects. Baboon #27, for example, may appear repeatedly in a range of presentations within the same conference. The prolific pig is not associated with this sort of economy of scarcity. Pigs are also understood as anatomically similar to humans: the heart, for instance, is just the right fit, and the ability to breed pigs of various sizes and weights means they bear the potential to generate parts for children as well as adults. Finally, pigs are categorized as farm rather than research animals, so they fall under radically different—that is, less stringent—regulations that dictate the nature of their care within the laboratory (Bayne 1999; Cooper et al. 2002; Niemann and Kues 2003).

The pig now stands at the forefront of xenotransplant research as the new potential "donor" that may one day alleviate or end the organ scarcity crisis. Baboons and, at times, rhesus macaques currently are featured not as sources of coveted organs but as the experimental precursors to human grafting. In other words, they are used as experimental recipients of porcine organs. As baboons and other simians recede to the sidelines, researchers celebrate their favoritism for the pig through an intriguing set of visual and rhetorical forms that together reveal an intimacy not shared with simians. This is especially evident when conference presentations on research findings include photographs or drawings of animals. Among the most striking differences is the distance maintained between humans and simians, whereas humans may be featured as literally embracing their pigs. As figures 3.2 and 3.6 reveal, the affection expressed by American researchers for the domesticated farm animal within the laboratory dates, at the very least, to thirty years ago. In figure 3.2, the young Robert Jarvik—now recognized as a significant pioneer in artificial heart research—is apparently attempting to kiss a calf as he tends to its dietary needs. In figure 3.6, the somewhat glassy-eyed head of the experimental calf AOPA is snuggled against that of a young female lab technician. Similar poses now characterize photos taken by researchers who work with hybrid miniature swine, and these images are regularly integrated into presentations on research at professional conferences.

Legal protocols most certainly play a part in shaping this striking distinction. As one postdoctoral student explained to me in 2004, researchers based in the United States warn their students to be careful about the way they employ visual images of their simian research subjects, concerned that photos

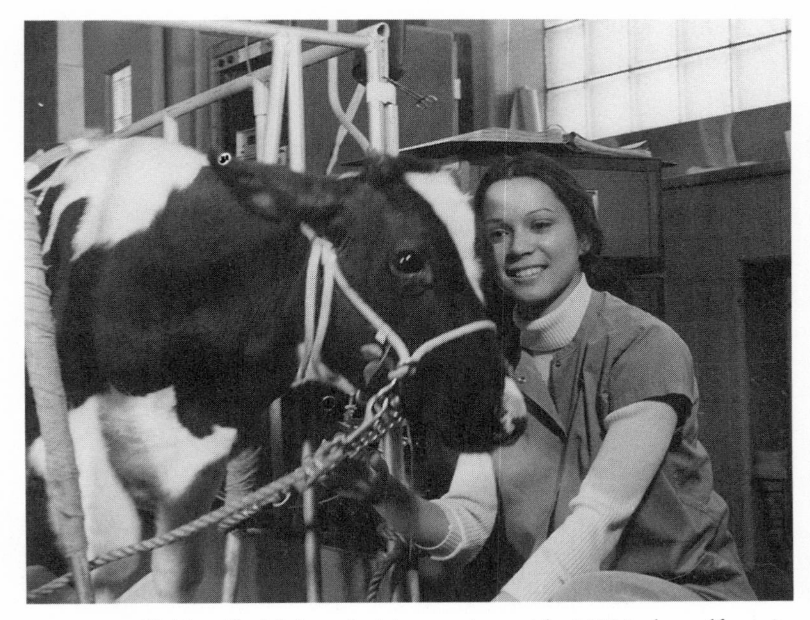

FIGURE 3.6. Unidentified lab technician posing with AOPA the calf; as its original caption explains, this "was taken on March 8, 1976, 118 days after his natural heart was replaced with an artificial heart. AOPA established a new record of 122 days." From a scrapbook compiled by Dr. Willem Kolff. *Special Collections, J. Willard Marriott Library, University of Utah.*

of individual animals may spark an investigation into the treatment of a specific creature. Yet other clues bear evidence to the fact that researchers' emotional investment differs between pigs and monkeys (although the same cannot be said for their handlers). For example, when photographs of baboons are featured in PowerPoint presentations, a solitary animal is shown, without exception, as seated calmly in a cage, preferably looking directly at us, and often either holding a banana or with a banana placed close by. This iconographic representation asserts that the animal is a caged research subject, yet is fed and thus well nourished. The food is also quintessential "monkey" food, serving to underscore that the animal is not human but a wild, though now somewhat tame creature. The presence of the cage also emphasizes, in dramatic fashion, how separate this "cousin" is from us. In other instances, where photos do not appear, illustrations inevitably feature the stylized and lanky outline of a gibbon or langur (with long arms, legs, and sometimes a tail), not the solid, stockier shape of other, squat Old World monkeys currently favored by researchers, namely, the rhesus macaque and baboon. We are to think of "baboon #27," for example, as a generic, research *monkey*.

Pig iconography is radically different, for these creatures are most often represented as endearing and Babe-like. Pigs featured in photos are almost exclusively cute little piglets; they are not shown in the uglier and more rotund form of the full-grown hog. (I must admit that I find the hybrid miniature pig to be unbelievably homely in its adult state, hairless and somewhat malformed in appearance.) During conferences researchers may also refer to an individual pig by name, especially if it is a breed sow that has helped establish a new genetic line (a practice reminiscent of Dolly the Sheep) (Franklin 2003).

At times pigs are pictured in their spacious pens, but the camera typically looks down on the animals through an open gate, not through the wire or posts surrounding them. Thus, as viewers, we again dominate the animal, but as a farmed and domesticated species, it generates no sense of harm or danger, as does the caged baboon. Another surprising tendency exhibited by researchers is that they may pose with and even hug their pigs. This may occur, too, in environments outside the research laboratory, as if the animal were a beloved pet or child. Among my favorite examples is a hybrid pig peeking over the back of a sofa. Yet another widely displayed image features a medical resident, in surgical garb, crouching next to a prized piggy, in the open door of a cage, an arm draped over the animal. Both subjects even appear to be smiling for the camera. Cuteness aside, this is a creature whose scientific value depends on its devaluation in quotidian contexts. Within labs where xenotransplantation defines a primary experimental concern, pigs emerge as extraordinary and beloved creatures.

THE DREAMS OF VENTURE CAPITAL

Within the realm of financial investment, however, scarcity heralds something quite different: the potential for entry-level investment strategies. As an interlude of sorts, I now wish to consider the relevance of venture capital by drawing from a 2000 volume of *Red Herring*, a financial magazine that specializes in technological innovations, and one that is popular among investors. In an article entitled "Machines for Living," Thomas Maeder and Philip Ross describe a scientific "battle to develop substitutes" for human organ replacement, marked by what they refer to as "mech" and "orgo," or mechanical and organic, alternatives.[4] As they insist, in terms of "mechabucks," in the end "both will prevail." The potential success of xenotransplantation is illustrated by a timeline of events that runs across the bottom of the magazine's pages like colorful, red-tinted ticker tape; the article itself focuses on the more peculiar or seemingly science fiction examples of experimental success. The narrative opens with the ominous statement, "Since the time of the Pharaohs, physicians have fashioned replacements for external parts of the body—peg legs, hook hands, glass eyes, false teeth, fake noses, wigs" with "the grafting of internal organs" as "the greater dream."

Maeder and Ross underscore the widespread use of all kinds of mechanical devices that are in place in approximately 20 million patients worldwide. Xenotransplantation thus emerges as only one of several possible solutions to the chronic organ shortage. As they explain, "Pigs, which resemble people more than we care to admit, may be the best candidates," with transgenic (or "cloned") piglets especially promising. Further, "The bold, the arrogant, and the optimistic would skip transplantation altogether and engineer artificial organs from scratch." The essay thus privileges "mech" as sexier (it includes several photos of mechanical devices but none of simians or pigs). From the authors' point of view, "orgo" is a messy, fleshy, and thus ultimately inferior model. The technologies that warrant the greatest attention at this point are those that can replicate human organ function in elegant mechanical fashion. Among my favorites is their description of the pancreas, "a nondescript bit of meat near the liver" responsible for keeping the body's blood sugar concentration in check. These marvelous "sugar daddies" (a wonderful pun, given the context) are a focus for much research involving everything from islet cell replacement to internally implanted mechanical devices. A bar graph entitled "Money Entrail" illustrates the great potential of both "mech" and "orgo" innovations; further, by 1999, "an estimated $305 billion was spent on introducing organic and mechanical enhancement to almost 25 million people worldwide." In the end, "mech" inventions clearly whet financial appetites; the "orgo" approach remains messier and thus riskier as far as investment capital is concerned. Ironically, the assumed inefficiency of "orgo" is rooted in the natural human body. At the very end of the essay, the "mammalian lung" is described as "perhaps one of nature's few engineering blunders" and the authors suggest that we would be better off with gills. One wonders if perhaps we should not simply convert (or submit) ourselves entirely to machines of human design. Needless to say, there is little concern voiced for the dangers of experimental procedures, for the quality of life of patients involved in research trials, or for the toll these technologies take on the lives of experimental research animals. The "mechabucks" stop here (Maeder and Ross 2002).

IMAGINING THE HYBRID BODY

Whereas the scarcity of human organs inspires venture capitalists to invest in alternative, experimental sources, for those awaiting transplants it is often a life-and-death issue. Transplant patients know all too well that their lives depend on the altruistic efforts of others, the vast majority of whom are strangers. They also know that the disparity between supply and demand grows each day, diminishing their chances of receiving a matching organ. Many will die waiting for one. As detailed in chapter 2, those fortunate enough to have experienced (and survived) a transplant often endure a heavy emotional burden shaped by the sense that their survival was made possible by another's death.

Organ recipients enthusiastically support experimental efforts to broaden the supply of transplantable parts or eliminate the need for human organs altogether. The majority, however, remain relatively uninformed of the details of current research, primarily because large-scale events or smaller workshops organized with their needs in mind more typically highlight breakthroughs in immunosuppression, or offer advice on such issues as post-transplant pregnancy, osteoporosis, or disability insurance and Medicaid coverage. Although xenotransplantation or heart implants may be featured in a plethora of PowerPoint presentations, these two topics are generally highlighted singly and in passing during rapid-fire overviews of transplant history. In short, recipients rarely, if ever, have the opportunity to hear from leading researchers on the intricacies of laboratory hybrid research.

When asked to consider experimental alternatives, recipients typically respond with joking behavior, silence, or abrupt changes of subject. These are not, however, indications of lack of concern. The more poignant moments in anthropological research may require deciphering the meaning of silence; and anthropologists have long recognized that particular forms of joking behavior are markers of social tensions and anxieties. I explore the significance of such reactions in detail elsewhere (Sharp in press). Briefly, though, recipients far more readily embrace the idea of mechanical devices than they do parts of porcine origin. They are already well aware of the dehumanizing effects of mechanical devices, for all have awakened from surgery to find themselves connected to a respirator (a device considered so irksome that attempts to yank out the tube are considered a clinical sign of recovery). The vast majority of kidney recipients previously endured hours of dialysis several times a week, and may have witnessed the deaths of others stationed nearby when machines have malfunctioned and attendants were busy elsewhere. Heart recipients, too, may have been among the legions of "pole pushers," waiting for an organ match alongside others on hospital cardiac units (Siebert 2004:169–72). Nevertheless, when discussing the future of transplant experimentation, most are not interested in receiving parts from animals.

Among the most striking aspects of these patients' answers to questions about animal and mechanical replacement options is the widespread sentiment that machines, unlike animals, are benign alternatives. Although they are certainly bound to break, they are perceived as bearing little potential to transform the self. Put another way, the machine makes no demands on recipients for new forms of sociality; unlike the pig or baboon, it is inert. With this, we come full circle in considering the social meanings associated with transplanted parts. As noted earlier in this work, organ recipients are frequently encouraged by professionals to imagine their organs as replaceable parts, their bodies as similar to automobiles, and transplantation as an elaborate form of repair. Organs, bodies, and transplant surgery are thus regularly "de-personified," so to speak. Attempts by recipients and kin to find one another, where intimacy is enabled by the sharing of blood, are aberrant and thus transgressive acts. Because sociality already looms so large in the recipient imagination, the transferred flesh of animals makes demands that may be too difficult to bear. Beyond the realm of companionship, there is no room within our society for human-animal couplings. Porcine and simian implants prove too threatening to the integrity of the body, the self, and society. If a person must lose their natal heart, let it be to a machine, not an undersized pig. No recipient I have encountered to date is willing to be debased by the beast.

ENDINGS AND NEW BEGINNINGS

An adage long asserted within anthropology is that we render the strange familiar and the familiar strange, a sentiment that is borne out by my analysis of simian subjects and transgenic pigs. Regardless of scientific celebrations, the transgenic pig remains a deeply troubling creature, its peculiar uniqueness rooted in its genetic potential. Already part human, it depends for its future on the ability to offer a steady supply of parts that will alter organ transfer as we currently know and understand it. Transgenesis defines a rich ground for analysis because it is fraught with both pragmatic and moral dilemmas that flag the values we assign to the whole, integral, human body. Can, or should, our bodies be melded with those of monkeys or pigs? What will become of our species if we allow such technological transgressions to breach the body's assumed "natural" boundaries? What social dangers lurk in such forms of scientific tampering?

As a means to conclude this three-part discussion of bodies, commodities, and biotechnologies, I wish to revisit one last time the sentimental quality of the new genetics. As I have illustrated over the course of this work, organ transplant recipients experience remarkable transformations by virtue of the fact that they harbor in their bodies fragments derived from other, now deceased people. This extraordinary development transforms a recipient's sense of self into a gestalt composed of ego merged with another. The new genetics insists on radical forms of intimacy with assumed baser creatures, animals we might one day embrace as being an even closer category of kin than their present categorization as distant, simian cousins allows. If the process of sociality already transforms the anonymous organ donor-as-stranger into a familiar intimate, what is the future potential of even stranger creatures?

There is "wisdom [in] repugnance," writes ethicist Leon Kass (Kass 1998), an adage many anthropologists would embrace, though not because it underscores danger in bioethical terms but because repugnance signals culturally encoded anxieties worthy of careful investigation. The beauty of the mechanical device is that it is fully inert, its origins rooted in the medical engineer's laboratory. But when organs are of flesh and blood, it may be impossible to deny their origins. The trouble with fleshy parts, then, is that they bear the weight of too much history and, thus, the inescapability of self-transformation. They blur those crucial distinctions we feel we must make between the familiar and the strange in the troublesome yet wondrous realm of futuristic transplant research.

EPILOGUE

The Future of the Body Transformed

Any anthropologist who wishes to track emergent practices along the cutting edge of transplant research faces a Sisyphean task. American biomedicine is, after all, driven in large part by a tireless desire to perfect its craft, and this is particularly pronounced where forms of surgical body repair are concerned. Trying to write of new findings is thus quite difficult, because, frankly, stasis is uncharacteristic of organ transfer in this country. The formal program of any professional conference confirms the vibrancy of a field marked by an exacting urge among specialists to tinker with and, thus, further develop their techniques. We may nevertheless speak in generalities as a means to characterize the nature of ongoing work: for instance, researchers' actions consistently focus on anticipating and preventing surgical complications, ensuring a better quality of life for transplant recipients, and, of course, identifying new sources of viable organs.

These prevailing concerns emerge readily, and are discussed openly, within a range of professional forums, including formal conference sessions, as well as less structured—yet equally important—debates that arise during question-and-answer periods or over cocktails and meals. Whereas, for instance, surgeons tout new techniques for heart valve repair, kidney extraction, or liver transection, procurement specialists explore how best to extend the "life" of cadaveric organs through careful donor management and subsequent preservation techniques they might employ as they transport precious parts. The hallways and exhibition halls of conferences are lined with temporary booths erected by biotechnology and pharmaceutical firms, whose representatives have lugged along an array of machines, rubbery mannequins, flashy posters, and other materials that promote everything from ergonomically designed surgical tools and portable pumps to the latest line of immunosuppressant medications.

In concluding this work, I wish at least to attempt to highlight a few activities drawn from the seemingly endless range of current innovations. My purpose here is twofold: to convey a sense of the energy (and urgency) invested in discovery and to underscore that medical personnel and bioengineers maintain no monopoly on innovation. As I have sought to show in the preceding three essays, the actions of lay participants likewise define a significant transformative presence within this specialized medical realm. In these remaining pages, then, I consider inventive responses mounted by recipients and donor kin and then turn to imaginative efforts set within the laboratory.

As I have illustrated throughout this work, it has become increasingly common among lay participants to subvert professional desires (and policies) as a means to reclaim rights to the donor body and its parts. Thus, whereas the dilemma of organ scarcity may trigger professional responses that ultimately depersonalize and, in turn, commodify the donor body, surviving kin and organ recipients regularly rehumanize supposed inanimate or soulless "spare parts" (Fox and Swazey 1992). Through this process, donors may be conceived of as reasserting their presence (albeit in a radically different form) among the living. Furthermore, the open-ended quality of death and associated grief work together not only to trigger such responses but also to awaken rather curious forms of sociality among strangers.

This book goes to press only a little over a year since I delivered the original lectures on which it is based. Yet within so brief a time, several remarkable shifts appear to be reshaping this *socialized* trajectory of organ transfer. For one, the language and imagery of war are increasingly pervasive. Whereas a decade ago, organ donors were regularly described as the "stars" of American transplantation, this collective term has been slowly eclipsed by "heroes." By 2006, however, within many quarters, organ donors as well as recipients are celebrated and marked rhetorically as national heroes (A. Jensen, personal communication, November 2005).[1] I believe that this discursive swell is due in part to the ongoing military campaign within Iraq, and the manner in which it weighs heavily on the American conscience. Still other factors are at work, too: as I argue elsewhere, activities central to organ procurement have been equated with the battlefield for several years (Sharp 2001). I cannot help but wonder if the now dominant trope of heroics might also be based on a growing, albeit unspoken, understanding of donors as medical martyrs who populate, in a rather liminal fashion, a highly technocratic realm where bodies and their parts are conceived of as so precious and so few.

Still another related, and emergent, theme concerns an even heavier emphasis on sociality, so that some procurement agencies now openly advertise that they coordinate "adopt a donor family" alongside "adopt a recipient" campaigns (A. Jensen, personal communication, November 2005). These paired practices are seemingly driven by the burgeoning sense that no one is alone in the world of organ transfer. Whether a person has given or received the "gift of life," they are entitled (and encouraged) to expand the boundaries of their web of kin in order to incorporate newfound "blood" relatives. This opportunity (or, potentially, obligation) extends to all involved parties, not simply to recipients who formerly, and alone, bore the weight of the "tyranny of the gift" (Fox and Swazey 1992).

A shift to laboratory settings reveals other intriguing examples of innovation, of a *technical* kind. Set alongside the competitive race between xenotransplantation and mechanical implants is yet a third entrant. Sometimes referred to as organogenesis, this area of specialization involves the bioengineering of tissues and even whole organs from cell cultures. Put simply, researchers are attempting to grow functioning organs from scratch. Several especially promising projects focus specifically on hollow organs and body parts (such as arteries, the ureter, bladder, uterus, vagina, and trachea). Cells are implanted on a scaffolding of biodegradable plastic, fashioned by the research team and replicating the shape of the intended organ. This seeded structure is then implanted within the host (or recipient) animal and allowed to grow in place. As cells establish themselves around the shape of the scaffold, the structure itself slowly dissolves, leaving a newly grown organ in place within the body.

Involved researchers thus far have been able to grow vascularized and innervated organs as rapidly as within six months. Successes have ranged from growing a vagina in a rabbit (and delivering a live pup) to crafting functioning bladders in pediatric patients who suffer from congenital disorders. This research is only partly dependent on access to stem cells (which need not be fetal in origin). Pediatric bladders, for example, are "tailor made" with cells harvested from the patient's original flawed organ and then reimplanted along with the new scaffold. Growing organs coveted specifically for routine forms of transplantation has proved more difficult. At least one team, however, has managed to generate a fleshy lump capable of excreting urine, as would a kidney (though malformed and reminiscent of raw hamburger in its appearance) (Atala 2005).

As is true with xenotransplantation and mechanical prototypes, organogenesis bears extraordinary promise for alleviating organ scarcity, especially if human parts could in fact be grown to order. What if, for instance, donor and recipient were the same patient, from whom cells could be cultured, grown externally, and then reimplanted? Such practices would render the cadaveric donor obsolete, and might easily circumvent the need for immunosuppressants. When taken together, these are simultaneously wondrous experiments that also generate intriguing questions about their social consequences.

As a medical anthropologist reflecting on these developments—whether social or technical in nature—I cannot help but return to the realm of sociality. How might freshly grown organs affect the seemingly radical, yet increasingly pervasive responses of donor kin and recipients? Would they one day be reinterpreted by professionals as terribly antiquated reactions to old-fashioned attempts to heal? Would patients' groups break up along lines separating those favoring what they might conceive of as a more natural, cadaveric origin for replacement parts, and those preferring to rely on medical expertise to grow their own, in vivo? My anthropological instincts anticipate that equally inventive strategies will emerge as a new generation of patients confronts the effects of radical experimental alternatives. As millennial medicine further refashions the self, these experimental subjects will most certainly generate a range of eclectic yet innovative ways for reimagining their origins within the medicalized frontier of bodies, commodities, and biotechnologies.

NOTES

INTRODUCTION

1. Throughout this work I frequently employ the terms "America" and "American" specifically in reference to U.S. contexts. My purpose is to avoid the awkward phrasing that characterizes such alternative expressions as "in/of the United States." That is, I do so out of convenience, and for poetic reasons, with no intention of neglecting the fact that the Americas contain other nations, too.

1. THE GOOD DEATH

This first chapter (as was the original lecture) is dedicated to Morton Klass (1927–2001), a beloved colleague, mentor, and friend.

1. The Organ Procurement and Transplantation Network (OPTN) maintains careful records of donation and transplant rates throughout the country, and it is the source for this and other related figures. OPTN is operated under contract by the the United Network for Organ Sharing (UNOS), a nonprofit organization, under the U.S. Department of Health and Human Services. See www.optn.org for past and current information.
2. As noted above, the liver of a living donor can be transected and, thus, shared with another patient in need; a person can also survive with a single lung, donating the other. Such transfers are relatively rare, however, because they endanger donors' lives; they most often involve organs culled from parents to save the lives of their children. A more complicated transfer is the "domino procedure," whereby a patient in need of a lung transplant (but who has a healthy heart) receives a pair of lungs and attached heart from a deceased donor. His or her diseased lungs are discarded, but the healthy natal heart is transferred to another heart recipient in need. Such circumstances allow for the donation of a heart from a living donor.

3. Again, see www.optn.org (accessed November 1, 2005).
4. Readers may wish to consult the following Web sites to view some examples: www.donatelife.net (Coalition on Donation), www.organdonor.gov (U.S. Department of Health and Human Services Web site on organ donation).
5. For other related visual examples, again see Sharp 2001:121–22.
6. When I visited this grove in mid-2004 there were two damaged stones devoid of plaques and five empty spots where, clearly, other trees had once stood.
7. I wish to thank Gregory Mann for first drawing my attention to these memorial sites in Fréjus.

2. BODY COMMODITIES

1. This figure was reported in October 2004 on the UNOS Web site at www.optn.org.
2. Long-standing legislation against prostitution most certainly informs this rhetoric, too.
3. See www.unos.org/helpSaveALife. In other contexts the figure is cited as being as high as 150 parts (see, for instance, Hogshire 1992).
4. For extended case studies see Sharp in press.
5. As is standard practice in anthropological research involving human subjects, the names of people and places have been altered in order to guard their privacy.

3. HUMAN, MONKEY, MACHINE

1. As this work went to press, the Coulson family reported that Miles, too, "is doing well . . . and acting just like a little boy should" (R. Dicks, Feb. 2006).
2. I have chosen to focus on Starzl because his personal career spans six decades, and because his writings, which are both clinical and autobiographical, offer a means to track the historical progress of xenografting from the mid-twentieth century to the present.
3. It was later learned that one of the patients was also infected with HIV. No match was available because at the time patients with HIV and hepatitis B were regularly excluded from transplant lists. In justifying their research to an internal ethics board, surgeons argued that both patients would soon die of their diseases with no hope for a human

organ graft. The argument of medical emergency is frequently invoked in life-threatening human experimental trials.

4. I would like to thank my father, Rodman Sharp, for drawing my attention to this article.

EPILOGUE

1. I am deeply indebted to Anja Jensen, from the Department of Anthropology, Faculty of Social Sciences, University of Copenhagen, for alerting me to several recent developments in the United States. Her keen eye for detail and novelty within the American context has proved especially helpful for me as I strive to update my own readings of organ transfer. I thank her for her time, and more importantly for insights drawn from her own field experiences while conducting research on donor families' experiences in the United States during the latter part of 2005.

REFERENCES CITED

Alexander, Caroline. 2004. "Across the River Styx." *The New Yorker*, October 25, 44–51.

American Medical Association (AMA). 2003. "AMA Testifies Before Congress on Organ Donation Motivation: Encourages Study of Financial Incentives." Press release.

Andrews, Lori and Dorothy Nelkin. 1998. "Whose Body Is It Anyway? Disputes Over Body Tissue in a Biotechnology Age." *Lancet* 351:53–57.

Antin, Joseph H., et al. 2001. "Establishment of Complete and Mixed Donor Chimerism After Allogeneic Lymphohematopoietic Transplantation: Recommendations from a Workshop at the 2001 Tandem Meetings." *Biology of Blood and Marrow Transplantation* 7:473–85.

Appadurai, Arjun. 1986a. "Introduction: Commodities and the Politics of Value." In *The Social Life of Things: Commodities in Cultural Perspective*, ed. A. Appadurai, 3–63. Cambridge: Cambridge University Press.

——, ed. 1986b. *The Social Life of Things: Commodities in Cultural Perspective*. Cambridge: Cambridge University Press.

Associated Press (AP). 2004. "Identities of Three Rabies Transplant Victims Released." *Houston Chronicle*, July 3, 37.

Atala, Anthony. 2005. "Tissue Engineered Artificial Organs: Current Concepts and Changing Trends" (ASAIO Lectureship). ASAIO (American Society for Artificial Internal Organs) 51st Annual Conference, "Enabling the Future Through Discovery and Innovation," Washington, DC.

Australian Broadcasting Corporation (ABC). 1996. "Artificial Hearts." *The Health Report*, Radio National Transcripts. Australia: ABC Radio National.

Bayne, Kathryn. 1999. "Developing Guidelines on the Care and Use of Animals." In *Xenotransplantation: Scientific Frontiers and Public Policy*, 105–10. New York: Annals of the New York Academy of Sciences, vol. 862.

Beidel, Deborah C. 1987. "Psychological Factors in Organ Transplantation." *Clinical Psychology Review* 7:677–94.

Berger, John. 1967. *A Fortunate Man: The Study of a Country Doctor*. New York: Pantheon.

BioTransplant, Inc. and Massachusetts General Hospital (MGH). 1999. "MGH and BioTransplant Scientists to Lead Discussions on Key Topics at International Xenotransplantation Association Conference." Press release, October 25, Nagoya, Japan on the PRNewswire.

Bourdieu, Pierre. 1994 (1977). "Structures, Habitus, Power: Basis for a Theory of Symbolic Power." In *Culture/Power/History: A Reader in Contemporary Social Theory*, ed. N. Dirks, G. Eley, and S. Ortner, 155–99. Princeton: Princeton University Press.

Centers for Disease Control and Prevention (CDC). 2004. "Investigation of Rabies Infections in Organ Donor and Transplant Recipients—Alabama, Arkansas, Oklahoma, and Texas, 2004." *MMWR* (dispatch), 1–3. Atlanta: Centers for Disease Control and Prevention.

CHFpatients.com. 2004. "Artificial Hearts." www.CHFpatients.com (accessed January 14).

Chillag, Kata. 1997. "Defining 'Normal': Representations of Patients' and Families' Experiences After Liver Transplantation." Ph.D. diss., University of Pittsburgh.

Cohen, Jeffrey Jerome, ed. 1996. *Monster Theory: Reading Culture*. Minneapolis and London: University of Minnesota Press.

Commissioners on Uniform State Laws (CUSL). 1980. "Uniform Determination of Death Act." Paper presented at Annual Conference of Commissioners on Uniform State Laws, Kauai, Hawaii.

——. 1988. "Uniform Anatomical Gift Act (1987)" (draft). Paper presented at Annual Conference, National Conference of Commissioners on Uniform State Laws, Newport Beach, CA.

Coombs, Robert H., et al. 1993. "Medical Slang and Its Functions." *Social Science and Medicine* 36 (8): 987–98.

Cooper, David K.C., Bernd Gollackner, and David H. Sachs. 2002. "Will the Pig Solve the Transplantation Backlog?" *Annual Review of Medicine* 53:133–47.

Csordas, Thomas J. 1994. "Introduction: The Body as Representation and Being-in-the-World." In *Embodiment and Experience: The Existential Ground of Culture and Self*, ed. T. Csordas, 1–24. Cambridge: Cambridge University Press.

Damasio, Antonio. 1999. *The Feeling of What Happens: Body and Emotion in the Making of Consciousness*. San Diego: Harcourt.

Das, Veena. 1997. "Language and Body: Transactions in the Construction of Pain." In *Social Suffering*, ed. A. Kleinman, V. Das, and M. Lock, 67–91. Berkeley: University of California Press.

Davis-Floyd, Robbie E. 1994. "The Technocratic Body: American Childbirth as Cultural Expression." *Social Science and Medicine* 38 (8): 1125–40.

Delmonico, Francis, L., et al. 2002. "Ethical Incentives—Not Payment—for Organ Donation." *New England Journal of Medicine* 346 (25): 2002–05.

di Leonardo, Micaela. 1998. *Exotics at Home: Anthropologists, Others, American Modernity*. Chicago: University of Chicago Press.

Dixon, Kathleen Marie. 1999. "Death and Remembrance: Addressing the Costs of Learning Anatomy Through the Memorialization of Donors." *The Journal of Clinical Ethics* 10 (4): 300–8.

Douglas, Mary. 1966. *Purity and Danger: An Analysis of Concepts of Pollution and Taboo*. New York: Praeger.

——. 1970. *Natural Symbols: Explorations in Cosmology*. New York: Pantheon.

Dunbar, Robin and Louise Barrett. 2000. *Cousins: Our Primate Relatives*. London: DK Publishers.

Feher, M., ed.. 1989. *Fragments for a History of the Human Body*, vol. 3. New York: Zone.

"First Heart Implant Dies." *The Courier-Journal* (Louisville, KY), December 1, 2001.

Fishman, Jay, David Sachs, and Rashid Shaikh, eds. 1998. *Xenotransplantation: Scientific Frontiers and Public Policy*. New York: Annals of the New York Academy of Sciences, vol. 862.

Flye, M. Wayne, ed. 1995. *Atlas of Organ Transplantation*. Philadelphia: Saunders.

Foucault, Michel. 1978. *The History of Sexuality, Vol. 1: An Introduction*. Trans. R. Hurley. New York: Pantheon.

Fox, Renée C. and Judith P. Swazey. 1992. *Spare Parts: Organ Replacement in Human Society*. Oxford: Oxford University Press.

——. 2004. "'He Knows That Machine Is His Mortality': Old and New Social and Cultural Patterns in the Clinical Trial of the AbioCor Artificial Heart." *Perspectives in Biology and Medicine* 47 (1): 74–99.

Franklin, Sarah. 1991. "Fetal Fascinations: New Medical Constructions of Fetal Personhood." In *Off-Centre: Feminism and Cultural Studies*, ed. C.L.S. Franklin and J. Stacey, 190–205. London: HarperCollins.

——. 2003. "Kinship, Genes and Cloning: Life After Dolly." In *Genetic Nature/Culture: Anthropology and Science Beyond the Two-Culture Divide*, ed. A. H. Goodman, D. Heath, and M. S. Lindee, 95–110. Berkeley: University of California.

Gould, S.J. 1981. "The Heart of Erminology: What Has an Abstruse Debate Over Evolutionary Logic Got to Do with Baby Fae?" *Natural History* 97:24.

Greenberg, Gary. 2003. "As Good as Dead: Is There Really Such a Thing as Brain Death?" *The New Yorker*, August 13, 36–41.

Hamilton, Anita. 2001. "Inventions of the Year/Your Health/AbioCor Artificial Heart." *TIME*, November 19.

Haraway, Donna J. 1991. *Simians, Cyborgs, and Women: The Reinvention of Nature*. New York: Routledge.

Harmon, Louise. 1998. *Fragments on the Deathwatch*. Boston: Beacon Press.

Harris, Marvin. 1985. *The Sacred Cow and the Abominable Pig: Riddles of Food and Culture*. New York: Simon and Schuster.

Helman, Cecil. 1988. "Dr. Frankenstein and the Industrial Body." *Anthropology Today* 4 (3): 14–16.

——. 1992. *The Body of Frankenstein's Monster: Essays in Myth and Medicine*. New York: Norton.

Helman, Christopher. 2001. "Charlotte's Goat." *Forbes Global*, February 19. http://www.forbes.com/global/2001/0219/061.html (accessed March 8, 2006).

Hogle, Linda F. 1995. "Tales from the Cryptic: Technology Meets Organism in the Living Cadaver." In *The Cyborg Handbook*, ed. C. H. Gray, 203–16. New York: Routledge.

Hogshire, Jim. 1992. *Sell Yourself to Science*. Port Townsend, WA: Loompanics Unlimited.

Houghton, Peter. 2001. *On Death and Not Dying*. London: Jessica Kingsley Publishers.

Institute of Medicine (IOM). 1997. *Approaching Death: Improving Care at the End of Life*. Washington, DC: National Academy Press.

Jankowski, Renee and Suzanne T. Ildstad. 1997. "Chimerism and Tolerance: From Freemartin Cattle and Neonatal Mice to Humans." *Human Immunology* 53:155–61.

Kass, Leon. 1998. "The Wisdom of Repugnance." In *The Ethics of Cloning Humans. A Reader*, ed. G. Pence, 13–37. Boulder: Rowman and Littlefield.

Kimbrell, Andrew. 1993. *The Human Body Shop: The Engineering and the Marketing of Life*. San Francisco: Harper and Row.

Leach, Edmund. 1964. "Anthropological Aspects of Language: Animal Categories and Verbal Abuse." In *New Directions in the Study of Language*, ed. E. Lenneberg, 23–64. Cambridge: MIT Press.

Leder, Drew. 1990. *The Absent Body*. Chicago: University of Chicago Press.

Lock, Margaret. 1996. "Death in Technological Time: Locating the End of Meaningful Life." *Medical Anthropology Quarterly* 10 (4): 575–600.

——. 1997. "Culture, Technology, and the New Death: Deadly Disputes in Japan and North America." *Culture* 17 (1–2): 27–48.

——. 2002. *Twice Dead: Organ Transplants and the Reinvention of Death*. Berkeley: University of California Press.

Machado, Nora. 1998. *Using the Bodies of the Dead. Legal, Ethical and Organisational Dimensions of Organ Transplantation*. Aldershot, Hampshire (UK): Ashgate.

Maeder, Thomas and Philip E. Ross. 2002. "Machines for Living." *Red Herring* 113:41–46.

Malinowski, Bronislaw. 1961 (1922). "Introduction: The Subject, Method and Scope of This Inquiry." In *Argonauts of the Western Pacific*, 1–25. Prospect Heights, IL: Waveland.

Marcus, George E., ed. 1995. *Technoscientific Imaginaries: Conversations, Profiles, and Memoirs*. Chicago: University of Chicago Press.

Marx, Karl. 1967. "The Fetishism of Commodities and the Secret Thereof." In *Capital*, vol. 1, part 1, section 4. New York: International Publishers.

Mauss, Marcel. 1967. *The Gift: Forms and Functions of Exchange in Archaic Societies*. Trans. I. Cunnison. New York: Norton.

——. 1973 (1935). "Techniques of the Body." *Economy and Society* 2:70–88.

Merleau-Ponty, M. 1962. *Phenomenology of Perception*. Trans. C. Smith. New York: Routledge.

Murphy, Robert F. 1987. *The Body Silent*. New York: Norton.

Murray, Thomas H. 1996. "Organ Vendors, Families, and the Gift of Life." In *Organ Transplantation: Meanings and Realities*, ed. S. J. Youngner, R. C. Fox, and L. J. O'Connell, 101–25. Madison: University of Wisconsin Press.

Nathan, Howard M. 2000. "Late Gov. Casey Made Sure Others Could Benefit From Organ Donors" (obituary). *Morning Call* (Allentown, PA), June 4, A11.

Niemann, H. and W.A. Kues. 2003. "Progress in Xenotransplantation Research Employing Transgenic Pigs." *Transplantationsmedizin* 15:3–14.

Nuland, Sherwin B. 1993. *How We Die: Reflections on Life's Final Chapter*. New York: Knopf.

Quaini, Federico, et al. 2002. "Chimerism of the Transplanted Heart." *The New England Journal of Medicine* 346:5–15.

Rabinow, Paul. 1992. "Artificiality and Enlightenment: From Sociobiology to Biosociality." In *Incorporations: Zone 6*, ed. J. Crary and S. Kwinter, 234–52. New York: Urzone.

——. 1996. *Making PCR: A Story of Biotechnology*. Chicago: University of Chicago Press.

Ragoné, Helena. 1994. *Surrogate Motherhood: Conception in the Heart*. Boulder: Westview Press.

——. 1996. "Chasing the Blood Tie: Surrogate Mothers, Adoptive Mothers and Fathers." *American Ethnologist* 23 (2): 352–65.

——. 1999. "The Gift of Life: Surrogate Motherhood, Gamete Donation and Constructions of Altruism." In *Transformative Motherhood: On Giving and Getting in a Consumer Culture*, ed. L. Layne, 65–87. New York: New York University Press.

Rapp, Rayna. 2000. *Testing Women, Testing the Fetus: The Social Impact of Amniocentesis in America*. New York: Routledge.

Richardson, Ruth. 1987. *Death, Dissection and the Destitute*. London: Routledge and Kegan Paul.

——. 1996. "Fearful Symmetry: Corpses for Anatomy, Organs for Transplantation?" In *Organ Transplantation: Meanings and Realities*, ed. S. J. Youngner, R. C. Fox, and L. J. O'Connell, 66–100. Madison: University of Wisconsin Press.

Roach, Mary. 2003. *Stiff: The Curious Life of Human Cadavers*. New York: Norton.

Rothman, David J. 1991. *Strangers at the Bedside: A History of How Law and Bioethics Transformed Medical Decision Making*. New York: Basic Books.

Sacks, Oliver. 1985. *The Man Who Mistook His Wife for a Hat and Other Clinical Tales*. New York: Simon and Schuster.

——. 1998. *A Leg to Stand On*. New York: Touchstone.

Scheper-Hughes, Nancy and Margaret Lock. 1987. "The Mindful Body: A Prolegomenon to Future Work in Medical Anthropology." *Medical Anthropology Quarterly* 1 (1): 6–41.

Schneider, David M. 1980 (1965). *American Kinship: A Cultural Account.* Chicago: University of Chicago Press.

——. 1984. *A Critique of the Study of Kinship.* Ann Arbor: University of Michigan Press.

Seale, Clive. 1998. *Constructing Death: The Sociology of Dying and Bereavement.* Cambridge: Cambridge University Press.

Sharp, Lesley A. 1994. "Organ Transplantation as a Transformative Experience: Anthropological Insights Into the Restructuring of the Self." *Medical Anthropology Quarterly* 9 (3): 357–89.

——. 1995. "Playboy Princely Spirits of Madagascar: Possession as Youthful Commentary and Social Critique." *Anthropological Quarterly* 68 (2): 75–88.

——. 1999. "A Medical Anthropologist's View of Posttransplant Compliance: The Underground Economy of Medical Survival." *Transplantation Proceedings* 31 (Suppl. 4A): 31S–33S.

——. 2000. "The Commodification of the Body and Its Parts." *Annual Review of Anthropology* 29:287–328.

——. 2001. "Commodified Kin: Death, Mourning, and Competing Claims on the Bodies of Organ Donors in the United States." *American Anthropologist* 103 (1): 1–21.

——. In press. *Strange Harvest: Organ Transplants, Denatured Bodies, and the Transformed Self.* Berkeley and Los Angeles: University of California Press.

Siebert, Charles. 2004. *A Man After His Own Heart: A True Story.* New York: Crown.

Starzl, Thomas E. 1992. *The Puzzle People: Memoirs of a Transplant Surgeon.* Pittsburgh and London: University of Pittsburgh Press.

Starzl, Thomas E., et al. 1993. "Baboon-to-Human Liver Transplantation." *The Lancet* 341 (8837): 65–71.

Stoller, Kenneth P. 1990. "Baby Fae: The Unlearned Lesson. Perspectives on Medical Research." Americans/Europeans/Japanese for Medical Advancement 2, www.curedisease.com (accessed December 2005).

Sturken, Marita. 1997. *Tangled Memories: The Vietnam War, the AIDS Epidemic, and the Politics of Remembering.* Berkeley: University of California Press.

United Network for Organ Sharing (UNOS). 2003. "Special National Donor Memorial Edition: Honoring America's Organ and Tissue Donors." *UNOS Update.*

——. 2004. "OPTN/UNOS Statement Regarding Rabies Transmission via

Organ Transplantation." Press release. Richmond, VA: United Network for Organ Sharing.

van den Bergh, J.C.J.M. and J.M. Holley. 2002. "An Environmental-Economic Assessment of Genetic Modification of Agricultural Crops." *Futures* 34:807–22.

Verdery, Katherine. 1999. *The Political Lives of Dead Bodies: Reburial and Postsocialist Change*. New York: Columbia University Press.

Webster, Donovan. 1996. *Aftermath: The Remnants of War. From Landmines to Chemical Warfare—The Devastating Effects of Modern Combat*. New York: Pantheon.

Youngner, Stuart J. 1990. "Organ Retrieval: Can We Ignore the Dark Side?" *Transplantation Proceedings* 22 (3): 1014–15.

INDEX